THE EMERGENCE OF ANIMALS
THE CAMBRIAN BREAKTHROUGH

captions on page iv

The Emergence of Animals

THE CAMBRIAN BREAKTHROUGH

MARK A. S. MCMENAMIN

AND

DIANNA L. SCHULTE MCMENAMIN

COLUMBIA UNIVERSITY PRESS

NEW YORK

Captions for frontispiece:

TOP
The Lower Cambrian trilobite *Dolerolenus zoppii* from southwestern Sardinia. Length of specimen approximately 3.5 cm. (Photograph courtesy Gian Luigi Pillola)

BOTTOM
The late Precambrian bedding-plane trace fossil *Harlaniella podolica* from the Chapel Island Formation, Fortune Head, Newfoundland. Scale bar is in millimeters. (Photograph courtesy Guy M. Narbonne)

Columbia University Press
New York   Oxford
Copyright © 1990 Columbia University Press
All rights reserved

Library of Congress Cataloging-in-Publication Data

McMenamin, Mark A.
  The emergence of animals : the Cambrian breakthrough / Mark A. S. McMenamin and Dianna L. Schulte McMenamin.
    p.   cm.
  Bibliography: p.
  Includes index.
  ISBN 0-231-06646-5.—ISBN 0-231-06647-3 (pbk.)
  1. Paleontology—Precambrian.   2. Paleontology—Cambrian.
3. Evolution.   I. McMenamin, Dianna L. Schulte.   II. Title.
QE724.M36   1989
560'.171—dc20
89-35459
CIP

Casebound editions of Columbia University Press books are Smyth-sewn and printed on permanent and durable acid-free paper
∞

Printed in the United States of America
c 10 9 8 7 6 5 4 3 2 1

CONTENTS

PREFACE

Guide and beacon who constantly passes over the infinite seas,
whose depths the great gods of heaven do not know,
your gleaming rays go down into the Pit,
and the monsters of the deep see your light.

HYMN OF SUN WORSHIP, NINEVEH (MESOPOTAMIA),
SECOND MILLENNIUM B.C.

A NCIENT MYTHS and modern science are united by the desire of humans to know their place in the universe, their role on the earth. Religion through the millennia has grappled with the question of why life exists on earth. Science has only in recent centuries allowed us to approach the question of how life appeared and developed on the earth.

The story of the earliest animals is as important to ecology as the Big Bang is to physics; it is the moment when all of the major elements of the earth's biota were, for the first time, united in a global ecosystem. This occurred during a transitional period of ecological crisis and resulted in a short-lived episode of extraordinarily rapid evolution. Despite mass extinctions, asteroid and cometary impacts, and shuffling of tectonic plates over the intervening 500 or so million years, the most important aspects of earth's ecology have remained virtually unchanged since the Cambrian boundary. The present makeup of the earth's biota cannot be understood except with reference to the singular event of ecological upheaval that took place at the end of the Precambrian.

The earth today is in a state of ecological crisis comparable in many ways to the turmoil occurring just before and during the Cambrian. The types of changes humans are causing on the earth's surface have not been matched in magnitude since the Precambrian-Cambrian transition. Whether or not our changes will endure depends on how long we and our machines can sustain our influence over earth's biota. The animals that emerged at the Cambrian boundary began an ecosystem that has been ascendant for a half billion years and has survived severe challenges.

This book is for anyone interested in the dynamics of life during profound and perhaps chaotic change, and for anyone interested in the types of ecosystems which can exist (and have existed) on the earth's surface.

WE GRATEFULLY acknowledge the scientific assistance of S. E. Alderman, S. M. Awramik, S. Bengtson, D. E. Briggs, M. Chen, P. Cloud, S. Conway Morris, J. W. Durham, W. Evitt, M. A. Fedonkin, A. G. Fischer, G. G. Gibson, M. Godchaux, J. Gray, K. Gross, L. B. Halstead, R. J. F. Jenkins, D. L. Kidder, J. L. Kirschvink, E. Landing, C. Lochman-Balk, Jin-Lu Lin, J. M. Morales-Ramirez, G. M. Narbonne, A. R. Palmer, D. L. Pierce, S. Rachootin, J. Roldan-Quintana, S. M. Rowland, A. Yu. Rozanov, B. Runnegar, A. Seilacher, P. W. Signor, J. H. Stewart, J. W. Valentine, J. C. G. Walker, M. R. Walter, E. Yochelson, students of the "The Origin of Animals" seminars at Mount Holyoke College. Despite their assistance, the people in the list above do not necessarily agree with the arguments presented in this book. We also thank A. Werner and S. Gruber for assistance with materials, and C. Reardon for making sure our preschooler felt loved while we were working on this project. Any errors are our own; please let us know if you find any.

Research included in this book has been supported by the American Chemical Society-Petroleum Research Fund, and a Presidential Young Investigator Grant #EAR 8857995 from the National Science Foundation to M.A.S.M.

THE EMERGENCE OF ANIMALS
THE CAMBRIAN BREAKTHROUGH

The Cambrian Explosion

THE HISTORY of life is not a continuum but rather is punctuated by episodes of great change. Many of these episodes of change are marked by the appearances and disappearances of unique life forms. The greatest "appearance" episode in earth history occurred at the beginning of the Cambrian Period (sometime between 500 and 600 million years ago), when abundant shelled animals began to fossilize and to become part of the rock record.

Compared to the vastness of geologic time, the appearance of shelled animals in the fossil record is quite sudden. The reason for this abrupt appearance of animals has been a subject of dispute among geologists for 150 years, with some scientists arguing that the first animals actually had a long period of unfossilized or "hidden" evolution, and others claiming that the fossil record of early animals should be read literally as an evolutionary explosion of new life forms. This explosion was once thought to mark the origin of life itself, and for this reason the stratigraphic horizon of animal emergence (known as the Precambrian-Cambrian boundary) has become the cardinal division in the geologic time scale (figure 1.1).

The main problem surrounding this boundary was first recognized many years ago when the research program of modern stratigraphic geology began. The controversy has changed character in the 150 years since this problem was first posed. But the central question, which remains unresolved, is: why did early animals wait until the Cambrian boundary to make their appearance?

Before 1930, few earth scientists believed that there was any good evidence for life older than the Cambrian. In 1931, Morley E. Wilson, after examining the extensive preCambrian rocks of the Canadian Shield, felt that stromatolites (fossil microbial colonies discussed later in this chapter) were "the only [preCambrian] forms the organic origin of which has not been seriously questioned" (Wilson 1931:124). Subsequent study has proven that, in addition to stromatolites and other fossils of microorganisms, multicellular life and even animals are present in strata deposited before the Cambrian. The key mystery, then, is no longer cast as the contrast between fossiliferous rocks of the Cambrian and lifeless underlying sediments. The central question now asks why there was an apparent explosion of animal diversity at the beginning of the Cambrian. All paleontologists today agree that undoubted animal fossils are known from tens of millions of years before the Cambrian. This being the case, why did animals delay so long before inheriting the earth?

The Precambrian fossil record is poor compared to that of the Cambrian and later. Was there a long history of hidden animal evolution before the Cambrian? Was there a major turnover in marine ecology during the Cambrian boundary interval? Or did a new, "improved" type of animal appear at this time? Was the origin of animals a lucky fluke, or was it more or less inevitable once certain environmental conditions were met? To consider these and other questions about the Cambrian explosion, we need to explore the "poor but deserving" late Precambrian and earliest Cambrian fossil record.

Traditionally, the biota (the sum total of living things) is divided into two categories, plants and animals. Plants are able to create their own food from sunlight using photosynthesis and simple chemical compounds provided by the atmosphere, water, and the soil. This feeding strategy is called autotrophy—literally, feeding oneself. Animals, on the other hand, are dependent on other organisms for food, a feeding strategy called heterotrophy. Many animals, unlike

FIGURE 1.1. The Geologic Time Scale, showing the Eras, Periods, and Epochs of the last 700 million years. Note that Riphean, Precambrian, and Proterozoic are not part of the formal Era/Period/Epoch time classification.

Era	Period	Epoch	Millions of Years Ago
	Quaternary	Holocene	
		Pleistocene	
Cenozoic	Neogene	Pliocene	1.8
		Miocene	
	Paleogene	Oligocene	24
		Eocene	
		Paleocene	
Mesozoic	Cretaceous	Late	65
		Early	
	Jurassic	Late	
		Middle	
		Early	
	Triassic	Late	
		Middle	
		Early	
Paleozoic	Permian	Late	245
		Early	
	Carboniferous	Late	
		Early	
	Devonian	Late	
		Middle	
		Early	
	Silurian	Late	
		Early	
	Ordovician	Late	
		Early	
	Cambrian	Late	
		Middle	
		Early	
"Late Precambrian" or Proterozoic	Vendian	Late	~550[a]
		Early	
	"Riphean"		~680–700

[a] The Precambrian-Cambrian boundary

3

most plants, move around their habitat seeking food resources scattered throughout the environment.

The traditional distinction between plants and animals has had difficulties ever since the invention of the microscope and the discovery of unicellular organisms or microbes. Cyanobacteria, also called blue-green algae, are common microbes and are among the simplest and toughest forms of life. They are familiar today as the green scum that forms in pet watering dishes. Their green color indicates the presence of chlorophyll, a pigment that captures the energy of sunlight and allows the cyanobacteria to create food in the process called photosynthesis. Since these microbes are feeding themselves with the help of sunlight, their feeding strategy is called photoautotrophy—literally, feeding oneself with light. The difficulty that microbes pose for the classical distinction between plants and animals is this—many microbes are both *autotrophic* (feed themselves with sunlight or simple, energy-rich molecules) and *heterotrophic* (gain food from other living things). Few familiar, large multicellular plants and animals are able to do this, but for many microbes it is commonplace.

Biologists now recognize five major categories or kingdoms of living things based on biochemistry and cell structure (as opposed to just two kingdoms based on color and capacity for movement). The five newer groups are: monerans, protists, fungi, plants, and animals (Margulis and Schwartz 1982). Monerans are simple unicellular organisms such as bacteria and cyanobacteria. Protists are larger, more complex, mostly unicellular organisms such as the amoeba. Fungi are unusual unicellular and acellular organisms ("acellular" refers to the fact that the bodies of large fungi are generally not subdivided into discrete cells). Fungi are primarily terrestrial (living on land) and mostly heterotrophic. Plants are multicellular organisms that (with a few exceptions) use photosynthesis to create their food. Plants also have rigid cell walls for support, an asset considering the stationary lifestyles of most plants.

Animals are multicellular heterotrophs dependent on other organisms for food; animals with complex organ systems are called metazoans. Most familiar types of animals are metazoans. Metazoans do not have rigid cell walls because these could interfere with a common animal characteristic—motility or mobility. As will be discussed later, multicellular animals can participate in autotrophy only with the aid of internal, symbiotic microbes. It is interesting to note that in terms of biomass (total amount of living matter), monerans and protists in the first two kingdoms vastly outweigh the bio-masses of the other three multicellular kingdoms taken together.

Animals, then, are multicellular organisms, often motile and without rigid cell walls, which require other organisms (living or dead) as food sources or providers. Alas, this is not a complete solution to the problems with the traditional definition of animals; some modern fungi could fit this definition. This classification problem is even more vexing for paleontologists than for biologists. There are many ambiguous fossils which have been interpreted as animals but which may in fact have belonged to some other kingdom of organisms.

Fortunately for the science of biostratigraphy, it does not matter what kingdom any particular fossil belongs to. For the stratigrapher trying to determine the age of a particular stratum, the most important factors are: (a) that the fossil discovered in the layer of interest be reasonably common, and (b) that it be restricted to a particular segment of geologic time.

Animal fossils allowed one of the most important scientific breakthroughs of the last century, the development of biostratigraphy and the "modern" stratigraphic research program. All groups of organisms inhabit the earth for finite periods of time, and once extinct, particular groups never reappear. The tragic irreversibility of extinction has a positive side for geologists practicing stratigraphy. Stratigraphers attempt to date rocks using a simple principle; when rocks form layers, the oldest rocks are on the bottom of the pile and the youngest are at the top. This principle of superposition is a powerful tool for determining the relative ages of adjacent rock layers, but it has limited usefulness, particularly when one tries to compare the ages of completely dissimilar stacks of sedimentary rocks (called sedimentary sequences) in different parts of the world. In the 1830s, early stratigraphers recognized that certain animal and plant fossils were restricted to specific intervals of earth history. Rocks that contained similar groups of fossils had similar ages, and a global sequence of sedimentary rock ages could be made by the comparison throughout the world of fossil biotas unique to a particular time. This global sequence is the familiar geologic time scale (figure 1.1). The major subdivisions of this scale (periods) have been in international use for over 100 years.

The oldest widely accepted period in the geologic time scale is the Cambrian Period. Until the 1930s, the Cambrian Period contained the oldest unambiguous evidence for life (Wilson 1931; Vidal

1984). From the completion of the subdivision of the geologic time scale into periods (1879) until the 1930s, the base of the Cambrian was where the fossil record ended. Rocks older than the Cambrian are still unceremoniously lumped together as the Precambrian. The Precambrian encompasses more than five-sixths of geologic time, and it seems neglectful to call this huge block of time by describing what it isn't—"pre-Cambrian." Some geologists have proposed dividing the geologic time scale into the Phanerozoic ("age of visible life") and PrePhanerozoic to describe the presence and absence of large animal fossils. But the base of the Phanerozoic unfortunately does not correspond to the base of the Cambrian (conventionally recognized by the appearance of shelly fossils), and the term Pre-Phanerozoic has only caught on among a few specialists in Pre-Phanerozoic paleontology. So the name Precambrian has stuck, and it is testimony both to the importance of fossils for unravelling earth history and to our relative ignorance of old rocks lacking visible fossils. Not surprisingly, the major division in the geologic time scale is often called the Precambrian-Cambrian boundary. Before 1930, this boundary was commonly understood to mean "fossils above, no fossils below." Our understanding of the Precambrian-Cambrian boundary has become more elaborate, although—as discussed earlier—some of the central unanswered questions concerning this boundary are still unresolved.

Before the concept of geological periods gained wide acceptance, the oldest fossils known were referred to as "Primordial." Many fossils that were first described as primordial would be placed today in the Cambrian Period. An early find of American "Primordial" fossils was made in 1843 by Asa Fitch in deformed strata in western Rensselaer County, New York, and described a year later by the early American geologist and obstetrician Ebenezer Emmons. Emmons was born in the Massachusetts section of the Berkshire-Taconic ranges, not far from our home institution of Mount Holyoke College. These fossils were the trilobites *Elliptocephala asaphoides* (figure 1.2) and *Atops trilineatus* from shales east of the Hudson River (Emmons 1847). Emmons believed that he had located the "Primordial fauna"—the opening chapter of life history on earth.

Emmons' views were met with disdain by emminent geologists of his day. Prominent geologists such as Sir Charles Lyell, widely acclaimed as the greatest geologist of his century, were unwilling to accept the antiquity of Emmons' fossils (Lyell 1845, 1849). In addition to disputing Emmons' claims for the great age of his fossils,

Lyell and others also denigrated Emmons' interpretation of the geo-logic structure of this part of eastern New York state (Schneer 1969). Emmons soon became aware of the hostility to his ideas, and he later remarked that "I do not know that I am indebted to any one for favors, or for suggestions. Indeed, nothing very flattering has ever been said, or published, respecting the views I have maintained on this subject" (Emmons 1856:vii).

Emmons' bitterness was well justified. In a particularly sordid episode in the history of geology, the greatest geologists of Emmons' day (including Lyell and the Harvard biologist/geologist, Louis Agas-siz) successfully twisted the dispute over Emmons' scientific ideas into a personal attack in court. The primary objective of this legal action was to discredit Emmons and to destroy his personal reputa-tion (Schneer 1978; Johnson 1982).

Emmons turned out to be correct about the age of the fossils, which makes it especially ironic that he suffered character assassi-nation for presenting his scientific views. Emmons had, in essence, correctly answered one of the major stratigraphic problems of his day. He had shown that there was indeed a substantial interval of strata with metazoan fossils occurring below the oldest (Silurian Period) fossiliferous interval yet reported (Emmons 1847:49). He had described some of the oldest fossils known at the time of their discovery. They came from a stratigraphic level or horizon which we now know to be quite close to the base of the Cambrian. Despite their age, there is nothing particularly special about these fossils.

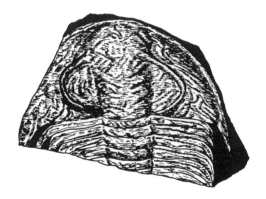

FIGURE 1.2. The first Early Cambrian fossil described from North America was the trilobite *Elliptocephala asaphoides.* (After Emmons 1847)

Elliptocephala and *Atops* are fully formed trilobites, not primitive ancestral forms. They are not unlike animal fossils that occur in much younger strata.

Charles Darwin was greatly distraught by the "explosion" of animal fossils at the Precambrian-Cambrian boundary. It posed a serious threat to his theory of evolution, which became generally available with the publication of *On the Origin of Species* in 1869. Darwin was strongly influenced by Lyell's (1830) vision of gradual, cyclic change throughout geologic time. Darwin's view of evolution was one of gradual change, in which one species slowly transformed into another in the fullness of geologic time. Darwin recognized that the the Precambrian-Cambrian boundary did not accord well with his gradualistic views. No animals were present before the Cambrian; then diverse, complex, fully formed groups of animals appeared after the boundary. In the sixth and last edition of his famous book, Darwin (1872:313) stated that "the case [for the abrupt appearance of Cambrian fossils] at present must remain inexplicable . . . and may be truly urged as a valid argument against the views [on evolution] here entertained."

Creationists still occasionally offer the Precambrian-Cambrian boundary problem as a fatal flaw for evolutionary theory (Gordon 1987). This is no longer a valid approach, however, because true animal fossils (soon to be discussed) are now known from sedimentary rocks which are much older that the base of (or lowest level of) the Cambrian. Creationists continue to trot out a once-respectable hypothesis (the sudden creation of Cambrian life) that was shown to be false by the middle part of this century.

Late nineteenth- and early twentieth-century paleontologists and geologists were unaware of these Precambrian animal fossils, so they were forced to come up with other explanations for the sudden appearance of Cambrian animals. Darwin, true to his preferences for gradual evolutionary change, reasoned that there was a gap in our knowledge of Cambrian ancestors, possibly due to a gap in the sedimentary record. Charles D. Walcott, discoverer of many important Cambrian fossils, followed Darwin's suggestion and formally postulated that the sudden appearance of fossils was due to a break in the recorded history of life. According to Walcott (1910), the earliest stages of animal evolution were not recorded as fossils because no sedimentary rocks were deposited and preserved during this time. He called this gap the Lipalian (from the Greek word for "lost") interval. A period of time not represented by sedimentary rock is often called a hiatus or unconformity by geologists, and the Lipalian interval was viewed as a worldwide, unbridgeable gap in the geological record. The Lipalian interval supported Darwin's gradualistic views, because it meant that the abrupt appearance of Cambrian animals was more apparent than real. Unfortunately for gradualism, this did not prove to be a viable solution to the Precambrian-Cambrian boundary problem. We now know of many sedimentary sequences that span the Precambrian-Cambrian boundary, but which lack the profound gap predicted by the Lipalian interval hypothesis.

The study of Precambrian paleontology began in 1858, when a collector for the Geological Survey of Canada found some curious specimens in very ancient metamorphic Precambrian rocks. These specimens were made of thin, alternating concentric layers of the calcium carbonate mineral calcite and the silicate mineral serpentine. Sir William Logan, director of the Canadian Geological Survey, thought that these banded specimens might be fossils. He was able to find better specimens near Ottowa in 1864. Logan brought them to J. William Dawson, principal of McGill University and the leading Canadian paleontologist. Dawson found what he thought were biologic structures in the calcite. Dawson identified these specimens as the skeletons of giant protists called foraminifera. Foraminifera can be very large by protistan standards, but these specimens were hundreds of times larger than any previously known foraminifer. Dawson named these objects "Eozoön canadense," literally, the dawn animal of Canada (O'Brien 1970, 1971). Not everyone accepted the biologic origin of "Eozoön," and Logan's pioneering study sparked a half century of controversy over whether there were any fossils at all in the Precambrian. Darwin was a staunch supporter of the biologic origin of "Eozoön" and he wrote that he had no doubt regarding its organic nature. This is not too surprising, considering the problems that Darwin's principal theory encountered if the Precambrian was lifeless. It was embarassing for the entire paleontological profession for there *not* to be any Precambrian life—all that strata with no demonstrable fossils! Dawson defended "Eozoön" to the end with what Gould (1980) calls "some of the most acerbic comments ever written by a scientist." Responding to Möbius (1879), a German critic of "Eozoön," Dawson (1879) remarked that Möbius did not have adequate geological background to evaluate the fossil properly, and charged that Möbius unfairly created a misleading view of "Eozoön" (O'Brien 1970). In spite of Dawson's deeply held desires, "Eozoön" proved to be inorganic, a product of metamorphosis found

9

only in rocks that have been altered by heat and pressure to the point that all fossils will have been destroyed (Hofmann 1982).

Unusual layered structures in sedimentary rocks, somewhat reminiscent of "Eozoön," were discovered by John Steele (1825) in Lester Park, New York State (in rocks now known to be Upper Cambrian in age). These structures, because of their size, shape, and internal layering, are very reminiscent of large cabbages which have been sliced in half. Several sedimentary horizons at Lester Park have abundant examples of these vertically stacked, hemispherical structures (figure 1.3). Steele did not recognize these structures as fossils, but the renowned New York paleontologist James Hall (1883) called these structures *Cryptozoön proliferum* (literally, "prolific hidden animal"). We now call these structures stromatolites. Stromatolites (figures 1.3 and 1.4) are curious fossils. They occur as regularly layered mounds of sediment (usually limestone or dolomite) whose layers curve upward and away from the substratum to form domes, cones, or branching columns. Stromatolites are both organic and sedimentary, having been built by the trapping and binding of sediment particles by communities of tangled threadlike monerans (mostly cyanobacteria). The tangled filaments, growing together, form a feltlike mat.

The formation of a stromatolite can be compared to the mound of dirt found within a clump of grass surrounded by barren, windswept earth. Not only does the grass clump prevent erosion of soil from immediately underneath it, it can actually trap windblown sediment because the velocity of the wind slows when it tries to blow through the clustered grass blades. When the wind velocity drops to a certain point, the soil particles that the wind had carried in suspension begin to fall, eventually adding to the mound of soil. Stromatolites form in a somewhat analogous fashion. When bottom currents sweep across the feltlike mesh of cyanobacterial filaments, the current slows and drops tiny, suspended sedimentary particles. With slow but steady growth, stromatolites add layer upon layer of trapped sediment particles to build the colony upwards and above the sea floor. The felty mat prevents erosion of the sediments stored underneath. A fossilized stromatolite records in its layers the growth and expansion of a moneran colony, and cannot be thought of as a fossil of a single organism, let alone an animal. The fossil history of stromatolites, however, has an important link to early animal evolution which will be discussed in chapter 8.

Field work by C. D. Walcott in the Grand Canyon (1899) and the

FIGURE 1.3. *Cryptozoön* specimens at Lester Park, near Saratoga, N.Y. Senior author (M.A.S.M.) is shown viewing these stromatolites. (Photograph by Mary E. Turner)

Belt Supergroup of Montana (1914) led to his recognition of the first Precambrian stromatolites. Walcott found microfossils in stromatolites from the Belt Supergroup which he correctly compared to modern photosynthesizing monerans. Walcott also described many supposed Precambrian animals. His description of Precambrian animals was largely in error, but as we shall see, he was by no means the last to make mistakes in the interpretation of supposed Precambrian animal fossils. Walcott's work established the importance of the Precambrian-Cambrian boundary as a discontinuity in the history of life, which was a big advance over the earlier ideas linking the origin of life to the "Primordial [read Cambrian] faunas." After Walcott's

FIGURE 1.4. An axial section through the stromatolite *Conophyton* from the Late Precambrian of Sonora, Mexico, showing the nested layers typical of stromatolites. Scale bar = 2 cm. (From Stewart et al. 1984)

death in 1927, Precambrian paleontology languished for a quarter of a century.

A renaissance in Precambrian paleontology began in the 1950s— discoveries and techniques developed then still guide current research activities. In 1954, S. A. Tyler and E. S. Barghoorn reported moneran microfossils in Precambrian cherts from near the Canadian shore of Lake Superior (Tyler and Barghoorn 1954). Chert, a microcrystalline form of silica, can beautifully preserve the delicate, microscopic structure of sea-floor monerans. Many Precambrian microbial fossil localities are now known from cherts, and some are from silicified stromatolites, confirming Walcott's inference about the organisms responsible for stromatolite formation.

The first conclusive proof of Precambrian animals came just before the 1950s with the discovery of organisms now referred to as the Ediacaran fauna.

Aliens Here on Earth?

Every important idea is simple.
LEO TOLSTOI

*For every complex problem there is a solution that is
simple, neat, and wrong.*
H. L. MENCKEN

B EFORE DISCUSSING the Ediacaran fauna, a distinction needs to
be made between the two major types of animal fossils—body
fossils and trace fossils. Body fossils are either actual parts of the
organism's body (such as a shell or a bone), or impressions of body
parts (even if the parts themselves have been dissolved away or
otherwise destroyed). The imprint of a feather or leaf or the external
surface of a shell are examples of body fossils. Occasionally, soft
body parts (such as the tentacles on a squid) can be preserved. As one
might expect, soft-bodied fossils are much rarer than shelly fossils,
because soft body parts are easily destroyed by decay and can be
fossilized only under exceptional preservation conditions.

Trace fossils are markings in the sediment (usually made while
the sediment was still soft) left by the feeding, traveling, or burrowing activities of animals. Familiar examples of trace fossils include
tracks and trails made by worms as they plow through sediment
looking for food and ingesting sediment. Animals making these traces

are usually eating organic matter in the sediment and are called deposit feeders. Footprints are another good example of trace fossils. The distinction between body and trace fossil can be blurred in some cases. Dinosaur footprints are known in which the scaly texture of the bottom of the animal's foot is preserved as part of the footprint. In a sense, this footprint fossil is both a trace fossil and a body fossil; it gives a record of the dinosaur's locomotion, and also gives a clear impression of the dinosaur's sole. Usually, however, it is difficult or impossible to exactly match trace fossils to the body fossils of the tracemaking organisms. Completely unrelated organisms can make trace fossils which are indistinguishable to paleontologists. Trace fossils are part of the fabric of the sediment, and therefore can be very resistant to destruction by metamorphism of the surrounding rock. Body fossils, on the other hand, are often destroyed by chemical reactions with the surrounding sediment. But body fossils are the only fossil type that can consistently give reliable information about the *identity* of the organism which left the remains.

In the 1920s, a peculiar body fossil was discovered in Germany and was claimed to be of Precambrian age. This fossil caused a minor sensation, first because it was unquestionably of biological origin, and second because it appeared to be a missing link between the Cambrian animals and their dark Precambrian past. Pompeckj (1927) called this fossil *Xenusion auerswaldae* (figure 2.1), and because it was apparently an incomplete specimen showing a body bearing numerous appendages or limbs, he believed that it was a fossil of a joint-legged animal. This was a satisfying conclusion, because joint-legged animals are an important group in the Cambrian. Trilobites, for instance, are abundant Cambrian joint-legged animals. Most modern joint-legged animals, such as insects and crabs, belong to the arthropod phylum (phylum is a rank of classification just below kingdom).

More difficult to interpret were a group of soft-body fossils found by German scientists in Namibia (formerly South-West Africa) in a sedimentary sequence called the Nama Group. Specimens were collected by P. Range and H. Schneiderhohn as early as 1908, but were not formally described in the scientific literature until the 1930s (Glaessner 1984). G. Gürich named one species *Rangea schneiderhohni* after its discoverers (Gürich 1930), and named *Pteridinium simplex* for its simple, fernlike shape (Gürich 1933; the Greek word for fern, genitive case, is *pteridos*). Both *Rangea* and *Pteridinium* (figure 2.2) are fern- or frond-shaped fossils. They both are accepted

today as important, genuinely Precambrian fossils, but unfortunately the early studies by Gürich were unablé to establish the age
of these fossils with certainty, because no undoubtedly Cambrian
fossils had been found stratigraphically above the beds with the soft-
bodied fossils. (Continued research has yet to uncover convincingly
Cambrian body fossils in the Nama sequence.)

Exciting discoveries by R. C. Sprigg in 1946 were used to establish
a definitive Precambrian age for frond fossils and other co-occurring
soft-bodied fossils. While doing some reconnaisance work as the
assistant government geologist of South Australia, Sprigg found
abundant soft-body fossils in the Ediacara Hills, a desolate, low range
in the desert some 600 km north of Adelaide, Australia (Glaessner
1984). The Ediacara locality subsequently became the most important locality in the world for Precambrian fossils of this type. These
soft-bodied organisms, presumed by many to have been metazoans,
are now referred to throughout the world as the Ediacaran fauna or

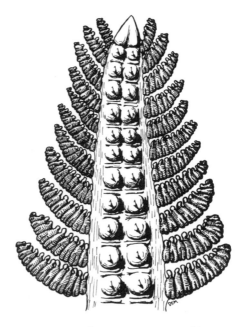

FIGURE 2.1. Reconstruction of *Xenusion auerswaldae,* an enigmatic Cambrian fossil originally interpreted as a Precambrian arthropod. Length of
frond approximately 7.5 cm. (From M. McMenamin 1986; used with permission of the Society of Economic Paleontologists and Mineralogists)

Ediacaran assemblage. The place name "Ediacara" is from the dialect of the 'Kujani, an aboriginal tribe. Loosely translated, it means "veinlike spring of water" (Jenkins 1984), an appropriate word origin considering that the Ediacaran fauna is the wellspring of all subsequent animal history.

Thousands of specimens have been recovered in the Ediacara Hills from platy quartzite slabs composed primarily of fine to medium sand grains. These rocks compose part of the Rawnsley Quartzite of the Pound Subgroup, a thick sequence of Precambrian marine sediments (Jenkins et al. 1983). The fossils are often large; discoidal and frond-shaped specimens sometimes exceed 1 meter in greatest dimension. Despite their relatively large size (for comparison, most Cambrian fossils are less than a few centimeters in maximum dimension, and no Cambrian animal is known to exceed 50 cm in length), the fossils of the Ediacara Hills are completely soft-bodied,

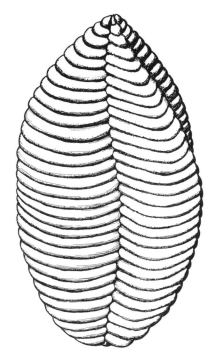

FIGURE 2.2. Reconstruction of a three-vaned frond fossil from North Carolina, probably belonging to the genus *Pteridinium*. Specimen approximately 7 cm long. (After Gibson et al. 1984)

17

with the possible exception of a few faintly preserved spicules (elongate spines) along the midline of one of the frond fossils.

Most of the Australian fossils are discoid impressions, nearly all of which have been interpreted as fossil jellyfish. This was Sprigg's first impression concerning the nature of the initial specimens collected; he thought he had found fossil "jellyfishes" (Sprigg 1947). Jellyfish belong to a large phylum of animals known as the Cnidaria. Cnidarians include, in addition to jellyfish, corals, sea pens, and sea anemones. All cnidarians share a basic trait—the presence of cnidae (also called nematocysts or stinging cells). These stinging cells are used for stunning and killing prey, and for defense against predators. These stinging cells are the bane of unwary bathers in jellyfish-infested waters.

Cnidae are not actually cells in the sense of having a nucleus with a cell membrane, but rather are greatly enlarged and specialized organelles. All plant and animal cells have organelles, tiny intracellular bodies that perform essential tasks such as photosynthesis (in chloroplasts) or respiration (in mitochondria, the "energy factories" of a cell). Most organelles are tiny, only a few fractions of a micron in greatest dimension. Cnidae are giants of the organelle world. They consist of a barbed, tightly coiled filament in a saclike body equipped with a hairlike trigger. When the hair trigger is contacted by a prey or foe, the filament is instantly discharged into the victim. Some filaments are over 1 millimeter (1000 microns) when extended.

The basic cnidarian body plan comes in two varieties: polypoid and medusoid. A sea anemone is a good example of a polypoid cnidarian. Its cylindrical, sac-shaped body is stuck to the substratum on one end and armed with a mouth and contractile tentacles on the other. Jellyfish are examples of the medusoid plan. The gut is similar in overall shape to that of the sea anemone, except that its body is inverted for swimming and is generally free from the substrate. The mouth, at the bottom center of the bell-shaped body, is connected by radial canals to the marginal parts of the bell. At the margin of the bell are fine tentacles loaded with cnidae, plus muscle fibers encircling the margin of the bell. Contraction of these muscles results in the characteristic pulsating swimming stroke of jellyfish.

There are three main shapes of Ediacaran fossils; circular impressions, frond fossils, and wormlike impressions. Circular impressions such as *Cyclomedusa* (figure 2.3), usually interpreted as medusoids, compose about 70 percent of the body fossil imprints in the Ediacara Hills collections. It is not at all clear, however, that these are all jellyfish impressions. At least one circular fossil, *Tribrachidium* (figure 2.4), is thought by most paleontologists to have no living relatives; it probably represents an extinct phylum.

Most other circular impressions of the Ediacaran fauna are classed as jellyfish or other cnidarians. *Skinnera*, for example, is a disc-shaped fossil with three inner "pouches" that Glaessner (1979) calls a jellyfish. The fossil *Mawsonites* (shown in color on the August 27, 1982, cover of *Science* magazine; Cloud and Glaessner 1982), although not considered by Glaessner (1979) to be undoubtedly a jellyfish, is assumed to represent the mouth-end (umbrellar surface) of a medusoid cnidarian.

Kimberella, an elongate bell-shaped fossil has been reconstucted by Jenkins (1984) as a cubozoan or box-jelly. Modern box-jellies are predators, fast and powerful swimmers. Jenkins' (1984) reconstruction of *Kimberella* has been called into question, however, because the fossil is triradially symmetric rather than having the fourfold radial symmetry required by the cubozoan hypothesis.

A number of different types of frond fossils are known from South Australia, including a few rare specimens of *Pteridinium*, which was

FIGURE 2.3. The discoidal fossil or "medusoid" *Cyclomedusa*, an Ediacaran body fossil. It has been interpreted by some as a jellyfish-like animal, despite the fact that, unlike in true jellyfish, concentric sculpture dominates the center of the organism, while radial elements dominate the margin. Diameter 6 cm.

first discovered in the Nama Group of Africa. Jenkins and Gehling (1978) consider the frond fossils of the Ediacara assemblage to be representatives of the modern pennatulaceans or sea pens. Pennatulaceans are colonial cnidarians with plumose colony outline that gives them their common name. The sea pen colony is fixed to the sea floor by a bulbous base; when disturbed, the plume part of the colony contracts and the creature seems to disappear. In a few specimens of the frond fossil *Charniodiscus,* known from England and Australia, the frond is associated with a circular, concentrically-ringed impression at the base of the frond. The frond seems to emanate from the center of the disk, leading to the reasonable interpretation that the disk was an attachment or holdfast for the frond. Glaessner (1959) first pointed out the striking similarities between these frond fossils and the body architecture of modern sea pens. Incidentally, the holdfast hypothesis may explain the nature of a large number of circular "jellyfish" impressions in the Ediacara assemblage; they could be holdfasts of which the upper frondose part of the colony was not preserved (Narbonne and Hofmann 1987).

The final major component of the Ediacaran fauna consists of ovate and leaf-shaped fossils that have been interpreted as fossil annelids (segmented worms) or flatworms. The organisms *Dickinsonia* and *Spriggina* (figures 2.5 and 2.6) are bilaterally symmetric and composed of numerous, elongate segments that seem to widen outward. No mouth or eyes are known from these fossils (despite an intensive search, using digital image analysis, by Kirschvink et al.

FIGURE 2.4. The enigmatic Precambrian fossil *Tribrachidium* from the Ediacaran fauna. Width approximately 2.5 cm.

FIGURE 2.5. A: *Dickinsonia costata*, a distinctive form from the Ediacaran fauna of Australia and northern Russia that reached up to one meter in length. This specimen was 4.6 cm in length. (After Runnegar 1982); B: another specimen of the same species from late Precambrian strata of the Flinders Range, South Australia. Greatest length of specimen 13.4 cm.

21

FIGURE 2.6. *Spriggina*, a wormlike form from the Ediacaran fauna. Maximum width of specimen 1.5 cm. (Copyright © 1989 by J. G. Gehling. All rights reserved)

1982), but one end of the elongate specimens of *Spriggina* is fused into a horseshoe shaped, headlike structure, and Runnegar (1982) has interpreted an elongate ridge running along the axis of *Dickinsonia* as a sediment-filled gut.

Nearly all current historical geology textbooks accept the above interpretations of the affinities of the Ediacaran fauna as essentially correct, and include a discussion of these fossils and their classification into familiar phyla such as Cnidaria and Annelida. There are, however, difficulties with the comfortable ensconcement of these fossils in modern phyla.

The first difficulty concerns the preservation of the soft bodies of the Ediacaran organisms. Jellyfish are uncommon but not unknown as fossils. Spectacular Jurassic jellyfish fossils are known from Germany, but these Jurassic jellyfish remains are in fine-grained lime-mud sediment, not the relatively coarse-grained, sandy sediments that preserve the Ediacara fossils. It seems strange that the delicate soft tissues of medusoids and worms were capable of being preserved in Precambrian sandstone in large numbers (thousands of Australian specimens are known). Such fossils are virtually unknown in sandstone deposits lain down later in earth history; modern jellyfish are more than 95 percent water by weight, and decay readily. Why didn't the Ediacaran animals decay away like their modern counterparts in similar (i.e., sandy nearshore) depositional environments? Glaessner (1984:48) addresses the problem of cnidarian impressions in sandstone with a brief account of his encounter with a stranded jellyfish. Glaessner stood on the jellyfish, and observed that the cnidarian was so tough that it could support his entire weight. The downward force of Glaessner's 80 kg body mass resulted in a "clear impression of its oral surface on the sand," but according to Glaessner there was "no damage to the medusa" (the jellyfish might disagree). Glaessner's experiment does attest to the strength of jellyfish tissue, but sheds little light on the burial conditions that entombed the Ediacaran creatures. No one was there to stand on the beach when a *Dickinsonia* was stranded.

An alternative to the "modern phyla" model for Ediacaran fauna affinites was offered by Adolf Seilacher in 1983. Seilacher is a West German paleontologist, well known for his innovative and sometimes controversial observations and studies of invertebrate paleontology.

The central point in Seilacher's critique of the conventional classification of Ediacaran fossils is this—Ediacaran organisms that preserve as soft-bodied fossils have very flat, almost ribbonlike bodies. This is not a body plan that is in accordance with classifying them as cnidarians and annelids, because in many cases it leaves no room for a gut, saclike or otherwise. Cnidarians, from large free-living individuals such as a jellyfish to branching colonies such as corals, are three dimensional rather than flat. Seilacher (1983, 1984) argues that the similarities between the Ediacaran creatures and modern cnidarians are superficial and that the conventional claims for close biological relationships between the two groups have not been examined critically.

Consider the sea pen classification proposed for Ediacaran frond fossils. The individual branchlets in a sea pen frond are separated, allowing water to pass between the branchlets and permitting each individual polypoid of the sea pen colony to capture food from the water streaming by. This mode of food capture is called filter feeding. In most Ediacaran fronds, the branchlets extending off of the main stalk of the frond are fused together. This does not mean that the organism could not filter feed, but filter feeding, if it occurred at all in an Ediacaran frond, must have been much different from the filter feeding in a modern sea pen. Without gaps between the segments of the frond, filter feeding of an Ediacaran frond would be a very inefficient process, at best.

Further, Seilacher (1984) maintains that there are fundamental differences between Ediacaran discoids and modern medusae. As outlined above, modern jellyfish medusae have radial gut structures in the middle of the "umbrella," and concentric structures representing annular muscle bands at the periphery of the medusae. The opposite is true for many of the discoid Precambrian fossils such as *Cyclomedusa* and *Mawsonites*, in which the sculpture in the central part of the disc is predominantly concentric, and the marginal sculpture is chiefly radial (Glaessner 1979).

Seilacher (1984) also takes a dim view of the hypothesis that *Dickinsonia* and *Spriggina* were worms. No definitive evidence has been found for eyes, mouths, anuses, locomotory appendages, or guts in these or related fossil organisms. These physical and behavioral traits are usually seen in segmented, annelid worms, as well as in unsegmented flatworms, but are not seen in the Ediacaran wormlike creatures.

The most radical part of Seilacher's (1984) critique of the modern phyla model is his suggestion that all of the Ediacaran soft-bodied fossils—disc, frond, and "worm" shapes—are related to each other and have little or no relationship to Cambrian or later animals. Seilacher (1985) calls these Precambrian soft-bodied fossils the "Vendozoa" (after the Soviet name for the last and only period of the Precambrian, the Vendian), an extinct taxon (or unit of classification) of ancient life perhaps comparable in rank to one of the five kingdoms such as fungi, animals, and plants discussed earlier. According to Seilacher, vendozoans went extinct at or near the Precambrian-Cambrian boundary, leaving few or no living descendants in a world overrun by "normal" animals (Seilacher 1984).

Seilacher (1984, 1985) sees vendozoans as variants of a single constructional principle; that of inflatable or pneumatic structures whose shapes are maintained by internal "quilting." In other words, the Ediacaran fauna consists of organisms that have a lot more in common (in constructional terms) with quilted air mattresses and inflatable rafts than they have with sea anemones and annelid worms. The pneumatic structure is a good way to make a body that holds its shape and yet is essentially two dimensional.

There is good evidence for this pneumatic architecture. Many Ediacaran fossils are apparently "deflated"; the upper membrane has collapsed downward between the more rigid baffles or walls between adjacent sections of the quilted pneumatic structure (figure 2.7). Like a flexible air mattress, Ediacaran soft-bodied creatures were capable of substantial expansion and contraction (Runnegar 1982) and perhaps did resemble inflatable rafts (except that they would have been inflated with water instead of air).

Seilacher's (1983) Vendozoa hypothesis caused quite a commotion when it was first presented at the annual Geological Society of America meeting in Indianapolis, and continues to be a topic of contention. Prominent articles discussing, and in general supporting, his views appeared in *Science* magazine (Lewin 1983) and *Natural History* (Gould 1986). There was no immediate response from sup-

FIGURE 2.7. Seilacher's (1984) interpretation of the structure of Ediacaran soft-bodied organisms. Left: inflated as in life. Right: deflated as in many fossil specimens. Note the relatively rigid vertical walls.

porters of the orthodox view of the Ediacaran fauna. Martin Glaessner and Mary Wade, long-standing proponents of the orthodox view of the Ediacaran fauna, have not yet published a response to the vendozoan hypothesis. Jenkins (1988) argues that arthropods, at least, must be present in the Ediacaran fauna, because of the presence of trace fossils that look as though they were formed by the legs of a Precambrian trilobite-like arthropod, but Jenkins has yet to publish photographs of these trace fossils.

The contrast between the conventional view and the Vendozoa hypothesis had not been fully resolved, but several paleontologists well-informed on the matter favor the orthodox viewpoint (Conway Morris 1987a, 1987b; Valentine and Erwin 1987). Seilacher himself takes the radical part of his Vendozoa hypothesis with a "grain of salt" and notes that the main function of this provocative idea is to stimulate research in the field of Precambrian paleontology and to free it from taxonomic preconceptions (Seilacher 1985). In a sense, Glaessner's and Seilacher's approaches are both extremes: the former sees the in the Ediacaran fauna mostly members of living phyla, and the latter sees mostly representatives of extinct phyla.

Which approach is closer to being correct? As Bengtson (1986) points out, the question cannot be answered with a simple verdict. Glaessner's philosophy of "shoehorning" Precambrian taxa into modern phlya is at least partly unjustified. Somewhat provocatively, Bengtson (1986) claims that instead of using "the present as the key to the past" (the fundamental geological mindset for interpreting past processes), Glaessner tends to use it as a "keyhole" to the past, letting our understanding of modern phyla narrow our field of vision to a small fraction of what it could have been, had we instead used it as a "key" to open the door to new interpretation. Indeed, some of the interpretations of the "Glaessner school" seem forced. Jenkins (1984; in his text, figure 2) reconstructs *Kimberella* with delicate cnidarian gonads, an interpretation that is in our opinion an overinterpretation. The vendozoan hypothesis merits serious attention, and we accept it as being closer to the truth than the conventional classification of Ediacaran fossils. At the very least, Seilacher has pointed out some important differences between the Ediacaran metazoans and their supposed cnidarian analogs.

We have much more information about the Ediacaran fauna than we did in the 1960s when much of the seminal taxonomic research was done by Martin Glaessner and Mary Wade. Nevertheless, it is not yet possible to unequivocally choose between such competing hypotheses as Seilacher's vendozoa hypothesis and the conventional view. Considering their impact on our understanding of early animal evolution, the unresolved questions of classification will continue to dominate discussion of the Ediacaran body fossils for some time to come.

Important as it is to know the biological affinities of the Ediacaran soft-bodied fossils, the Vendozoan controversy raises an even more important issue concerning these organisms. Many of the Ediacaran creatures were laterally compressed organisms, although a number of them were sac- or bag-shaped. Seilacher (1984, 1985) recognizes that flattened body shapes maximize surface area for the takeup of oxygen and food dissolved in seawater, and perhaps also for the absorption of light. "Normal" metazoan animals generally have plump, more or less cylindrical, bodies. For very small, thin skinned animals, cells near the body surface can get oxygen and expel waste by simple diffusion across the cell surface membranes. Waste products such as carbon dioxide will be supersaturated inside of the animal's body, and will tend to migrate out of its cells and into the open environment. The reverse is true for oxygen; it will tend to migrate into the cells because its concentration is greater on the outside than on the inside of an oxygen-respiring animal. Animals such as frogs and salamanders are able to respire (at least in part) in this way. But for most large, cylindrical animals, diffusion respiration will not work because diffusion is ineffective for cells buried deep within the animal's body. This is a consequence of the fact that as an animal increases its size, its total volume outstrips its surface area by a large margin.

When an animal grows larger without changing its shape, every linear dimension (length, for instance) will increase. But the surface area of the animal increases faster than the increase in linear dimension, and the volume of the animal increases faster still. More precisely, if the length of an animal is doubled, its surface area increases by a factor of four, and its volume goes up by a whopping factor of eight! The new animal twice as long has eight times as much body volume.

With such dramatic increases in volume, how can a "thick" organism care for the nutrient needs of the cells buried deep inside its body? A large, cylindrical animal is clearly too thick for diffusion to work as a means of respiration. We metazoans have developed intricate systems of pipework and tubing to deliver nutrient and waste removal services to interior cells. Circulatory systems, digestive tracts, gills, and lungs are all solutions to the problems associated with volume increase. But this is not the only possible solution for large organisms. *Dickinsonia,* up to a meter in diameter, was flat and thin enough to use its expanded body surface for respiration and waste removal (Runnegar 1982), thereby avoiding reliance on complex internal plumbing.

A number of living organisms survive in this way. Some modern protists, up to 38 millimeters in length, are able to feed without "eating" anything and without photosynthesizing. These foraminifera simply absorb nutrients dissolved in sea water (Delaca et al. 1981). Deep-sea worms, clams and other organisms living near mid-ocean hydrothermal vents absorb hydrogen sulfide (usually toxic to "normal" animals) from the vents, and with the help of internal, symbiotic bacteria, convert the sulfide into food. This arrangement between bacteria and metazoans is called chemosymbiosis. Photosymbiosis, an arrangement in which interal "gardens" of photosynthetic monerans live within the tissues of a metazoan host and provide it with food, is utilized by a number of modern animals, including corals and some clams. These "solar-powered" animals are proving to be more abundant in marine communities than previously thought (Rudman 1987). A beautiful sea slug (or nudibranch) called the blue dragon *(Pteraeolidia ianthina),* is a one-time-only predator of coral polyps. Early in its life, the young blue dragon eats a coral polyp. But instead of digesting its victim's internal photosymbionts (called zooxanthellae), the sea slug cultivates them and allows them to multiply within its own body. The blue dragon is apparently able to survive entirely on the excess foodstuffs made by the monerans, since adult blue dragons do not eat coral. The body of the blue dragon is covered with tubular outgrowths, called cerata. These cerata, filled with algae, form fanlike arrangements that optimize the light-gathering ability of the slug's body, and also give the slug its characteristic color (Rudman 1987). The adult blue dragon looks something like a sprig of blue spruce.

There is every reason to believe that some of the unorthodox feeding strategies noted above—passive nutrient uptake, chemosymbiosis, photosymbiosis—were employed by members of the Ediacaran fauna. A. G. Fischer (1965) was first to suggest that the flattened shapes of the Ediacaran fossils would have been conducive to the harboring of photosymbiont tenants. We recently asked Fischer to comment on his idea, now over two decades old:

Yes, so far as I know that thought of photosymbiont associations was original with me—not that that seems so important.

I find it hard to think of primitive things like that getting so big without some such mechanism—and I had no idea then of how big *Dickinsonia* actually gets or got (written communication, 1988).

Seilacher (1984) added that body shapes with high surface area are also good for absorption of simple compounds such as hydrogen sulfide, needed for one type of chemosymbiosis. A sea-floor "pancake" like *Dickinsonia* may have absorbed nutrients directly from sea water through its enhanced surface area, or alternatively, may have lived with an internal "garden" of symbiotic, photosynthetic, food-producing monerans or protists. Flatness of body is not, of course, a requirement for these feeding strategies. A newly named fig-shaped Ediacaran fossil *Inaria karli* (figure 2.8) was originally

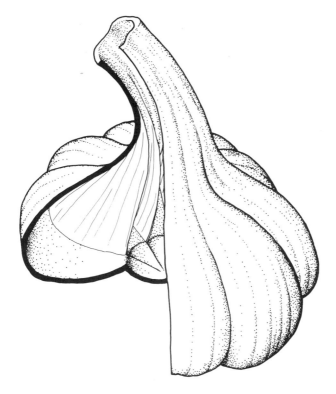

FIGURE 2.8. *Inaria karli* has a sac-shaped, instead of flat, body plan. This figure shows a cutaway of a mature specimen in life position. Greatest width of specimen about 7.5 cm. (After Gehling 1988)

quite three-dimensional. Gehling (1988) suggests that *Inaria* possessed a bag-shaped body plan designed for photosymbionts, and may have acted like a "respiring culture-chamber." Strategies such as these may have rendered these organisms independent of external, digestible, living sources of food. The ecological implications of such strategies will be considered in the chapter 7 discussion of the "Garden of Ediacara."

III

The Mudeaters

Test everything; retain what is good.
1 THESSALONIANS 12:21

THE WORST problem in the search for the oldest animal fossils is mistaken identity. Sedimentary rocks are replete with irregular structures and small scale disturbances or interruptions of the horizontal bedding or layering. Some of these disturbances are caused by organisms, but many are not. If these disturbances are formed at the same time that the sediment was deposited, they are called primary sedimentary structures. Trace fossils are primary sedimentary structures formed by the burrowing and feeding activities of animals that lived on or in the soft sediment (or, occasionally, hardened sediment).

Usually a well-preserved and well-formed trace fossil is unquestionably biologic in origin, and all paleontologists would agree that the trace was formed by an animal. Yet it can be difficult to define precisely what it is about a trace fossil that makes it convincingly biogenic (formed by life). It is like the statement by an American judge regarding his identification of pornography; he couldn't define precisely what pornography was, but he knew it when he saw it. Regularity of structure, a shape suggesting motion through mud by an animal, and the implausibility of forming a structure solely by physical processes are all criteria used to establish the biologic nature of trace fossils. As you might expect, there are many cases where the true identity of a trace-fossillike structure is not clear.

A sedimentary structure that resembles, but is in fact not, a trace fossil (or a body fossil, for that matter) is called a pseudofossil. Pseudofossils have plagued the study of Precambrian paleontology because many inorganic sediment disturbances look deceptively like fossils. While out in the field, earth scientists always hope their labors will be rewarded by a crucial discovery (such as a find of the oldest animal fossil) and by the professional acclaim that would attend such a discovery. Paleontologists have occasionally made the mistake that A. H. Knoll (1986) calls "highgrading" on the outcrop, passing over many ordinary, inorganic sedimentary structures in Precambrian rocks but collecting and publishing accounts of those that look most like fossils.

Figure 3.1 is a histogram showing the frequency of formally published goofs (cases of mistaken identity) in the search for the oldest animal fossils. As the histogram shows, there are three peaks in the data, one during the interval 1860–1870, a second during 1890–1900 and a third in the interval 1960–1970. The first peak includes *Eozoön* and other Precambrian "fossils" described by Dawson. The second peak is a result of C. D. Walcott's error-prone work in the identification of Precambrian animal fossils. We can only wonder why Walcott was so consistently wrong whenever he dealt with Precambrian animals. Perhaps he was striving to fulfill the predictions of Darwin's gradualistic views. Walcott *expected* to find Precambrian animal fossils, a bias which may account for many of his misidentifications.

The third peak in figure 3.1 is a result of burgeoning research in Precambrian geology beginning in the early 1960s, plus a spate of unfettered enthusiasm for attempts to find the most ancient animal fossil. Seilacher recalls (personal communication, 1987) a presentation given in the early 1960s at a formal geologic meeting. The speaker claimed to have found the oldest evidence for fossil metazoa, and showed the audience photographs of the most convincing specimens. After the presentation, several scientists in the audience expressed grave reservations about the putative biologic origin of these structures. A vote was immediately taken to see how many people in the audience were convinced of the biologic nature of the specimens. The vote was overwhelmingly in favor of the author's conclusions. Seilacher, who was one of the skeptics (and later proved correct), was so dismayed by this incident that it was twenty years before he again worked seriously with Precambrian animal fossils.

As seen in figure 3.1, the number of mistakenly described Precambrian pseudofossils has fallen off dramatically since the peak in the early 1960s. The drop-off in the numbers of cases of mistaken identity can be attributed, perhaps in large part, to the unflagging skepticism of Preston Cloud, a geologist at the University of California at Santa Barbara. Cloud published several articles (1968, 1973) attacking the uncritical approach taken by many of his colleagues toward the identification of Precambrian animals, particularly those older than the beds containing Ediacaran fossils.

I (M.A.S.M.) have added my own blunder to those tabulated in figure 3.1. As a graduate student in 1982, while doing field work in the Clemente Formation of the Cerros de la Ciénega of northern Sonora, Mexico, I came across some markings in the rock that looked suspiciously biogenic (figure 3.2). The rock, a red-colored sandy silt-

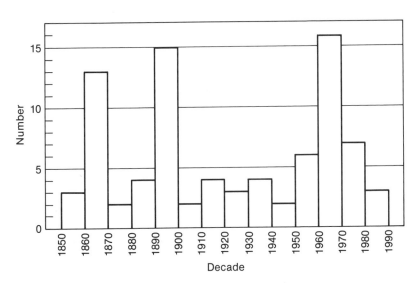

FIGURE 3.1. Number of spurious Precambrian animal fossil descriptions plotted against year. The three highest peaks in the data primarily result from the work of Dawson (1860–1870), Walcott (1890–1900), and various overenthusiastic recent workers (1960–1970).

stone, contained four elongate lobe-shaped objects, each with concentric U-shaped ridges at the end of the lobe. The U-shaped ridges looked a lot like the concentric layers of sediment that form in many types of animal-built burrows. I immediately showed the specimens to the leader of the field expedition I was on. He agreed that they looked very much like trace fossils. When I returned from the desert, I showed the objects to Preston Cloud. Cloud, who by 1982 had acquired a ferocious reputation as the premier debunker of Precambrian "fossils," agreed that they could be biologic. With youthful enthusiasm, I published a photograph of these objects (along with other discoveries from Mexico), describing them as "probable metazoan traces" (M. McMenamin et al. 1983). After publication, I returned to Sonora and attempted to find more specimens of convincing trace fossils in the Clemente Formation, the Precambrian rock unit that had produced these specimens, but without success. I showed the specimens again to Cloud. He still felt that they could be biologic, although he was bothered by the fact that all four of the lobes

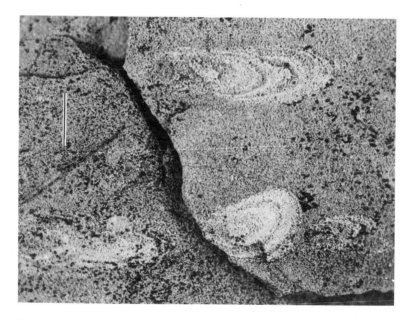

FIGURE 3.2. Precambrian psuedofossils, from the Clemente Formation, northwestern Sonora, Mexico. Scale bar = 1 cm. (From McMenamin et al. 1983)

were oriented in approximately the same direction. This made him suspicious, but he wasn't sure why. This later proved to be an astute observation. In 1984, while I was completing my doctoral thesis, the Australian paleontologist Malcolm R. Walter visited Santa Barbara. I showed him the enigmatic structures, and he immediately came up with a way to form these lobes without invoking animal activity. Walter has seen flow structures, resembling tiny lava flows off of the flanks of volcanoes, forming in association with sediment fluid-escape cones, also called sand volcanos. Sand volcanos can form when water-saturated sediment is exposed to air, and then disturbed by compactional forces or jostled by earthquakes. When this occurs, the sediment settles and forces water to move upwards. The water will sometimes follow a cylindrical conduit, roughly resembling the vent of an igneous volcano. Sediment entrained in the water stream will be deposited where the dewatering flow meets the air, and can be deposited in a broad-sediment cone, or sand volcano. These sand volcanos are often only a few centimeters in diameter, much smaller than their igneous counterparts. At the center of the sand mound is a small collapse pit, which looks like the vent, or caldera, in an igneous volcano.

When small sand volcanos are preserved in ancient sediments, they are called pit-and-mound structures. Sometimes a particularly fluid slurry is ejected from the sand volcano vent. This slurry can flow down and beyond the flanks of the sand volcano, forming a lobe of fluidized sediment that can settle to create a sedimentary structure that looks very much like a trace fossil with backfilled layers (figure 3.3). Ancient sand volcanos can be recognized in Precambrian sediments of Australia (Walter 1972). Walter (1976) has described similar structures forming today at Shoshone Geyser Basin, Yellowstone National Park. Pit-and-mound structures (figure 3.4) are known from the same locality in the Clemente Formation that yielded the trace-fossillike structures. I am now convinced that these objects are pseudofossils (M. McMenamin 1984), and I plan to be more skeptical in the future when I see Precambrian fossillike objects that are from rocks older than the Ediacaran soft-bodied fossils. Careful attention to the mode of formation of inorganic sedimentary structures can help one avoid the misidentification of pseudofossils.

Even with intense study by a number of different investigators, there are some Precambrian structures that remain possibly biologic and genuinely enigmatic. These unidentified fossillike objects (UFLOs) are continuing sources of dispute. Body fossils can be mimicked by

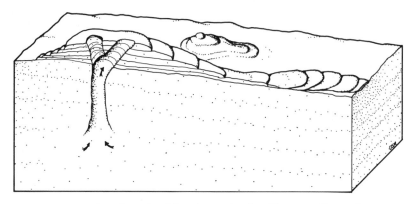

FIGURE 3.3. Inferred mode of formation for the Clemente Formation pseudofossils. Liquified sediment flowed upward and outward through a small
sand volcano. Width of sketch approximately 10 cm.

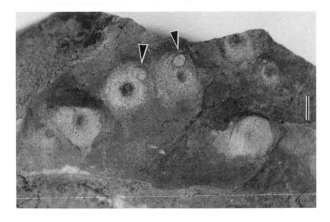

FIGURE 3.4. Sand volcanos, some with blisters (indicated by arrows), from
the Clemente Formation, northwestern Sonora, Mexico. Scale bar = 0.5 cm.

sedimentary structures of mechanical origin. Circular "medusoids"
have been recently reinterpreted as gas pits (Sun 1986) and very
ancient "frond-fossils" may be inorganic ice tracks, formed by the
scraping of a block of ice across sediments (Jenkins 1986). The UFLO
called *"Brooksella" canyonensis* is perhaps the most famous of these.
Described in 1941 by R. S. Bassler, this object (see Häntzschel 1975;
his figure 89) consists of a set of radial, overlapping petallike lobes
that look something like the remains of a daisy that has had its
yellow center removed. *"Brooksella" canyonensis* has been var-

iously interpreted as a jellyfish body fossil (Bassler 1941), a pseudofossil formed by gas bubble escape through soft sediment (Cloud 1968, 1973), and a starlike trace fossil, possibly a feeding burrow (Seilacher 1956). Kaufmann and Fursich (1983) support the trace fossil interpretation because of the presence of curved layers in one of the lobes which they interpret as evidence for metazoan burrowing activity. As shown in the discussion above, however, such layering does not rule out formation by an inorganic process.

Convincing trace fossils are known from the late Precambrian, sometimes in association with the soft bodied Ediacaran fossils (Glaessner 1969). These trace fossils are generally simpler, less common, and less diverse than Cambrian trace fossils. There is a significant difference in the complexity and depth of burrowing between Cambrian and Precambrian trace fossils, and it has been argued that the changeover from simple trace fossils to more complex types of traces occurred at more or less the same time as the Cambrian explosion, the first appearance of abundant Cambrian shelly fossils. In a paper that now seems surprisingly ahead of its time, Seilacher (1956) was first to point out the transition in the trace fossil record that occurred across the Precambrian-Cambrian boundary.

Marine trace fossils are generally of three varieties: locomotion traces, deposit feeding traces, and dwelling traces. The three types can and do grade into each other, but in most cases any particular trace fossil is more of one type than of another. Deposit feeding, for instance, nearly always has some component of locomotion because the animal needs to move around to find fresh, unmined organic deposits in the sediment for food. An example of a pure locomotion trace would be scurry marks made by a many-legged arthropod as it moves across a stiff mud, leaving claw marks behind.

Dwelling burrows are actually a form of skeleton made of sediment. The animal has excavated a cavity in the substrate, and this hole or tube in the mud serves as "home base" from which the resident can mine sediment or capture food particles suspended in the water with a filtering device. This type of burrow is also useful for avoiding predators that cannot burrow into the sediment.

Nearly all of the convincing trace fossils from the Precambrian are of the deposit feeding and locomotion varieties. Most of these are simple, tubular burrows of an animal that once moved through the soft (but now cemented into rock) sediment in a more-or-less unidirectional or slightly meandering fashion (Crimes 1987). These types of traces are commonly formed by many different types of unrelated, wormlike organisms (including some annelids or true worms), so it is impossible to learn much about the identity of the tracemaker from these simple traces. Sometimes these traces show evidence of peristalsis, or motion by rhythmic contraction of circular muscle bands along the length of the animal's tubular body. Earthworms move by peristalsis, and food is moved down your esophagus by a similar sequence of muscle contractions.

Few Precambrian animals had learned the "trick" of excavating a home from the sediment. Deep vertical, cylindrical burrows are very rare in Precambrian sediments. Organisms that make these kinds of burrows are often sessile, which means that they spend large amounts of their lives staying in one place. The simplest dwelling trace fossil of this type is called *Skolithos* (figure 3.5). *Skolithos* is an unbranched vertical, tilted, or curving cylindrical burrow. The oldest deep (greater than 1 cm in depth), undoubted burrows are Precam-

FIGURE 3.5. Vertical burrows called *Skolithos*, from the Lower Cambrian Proveedora Quartzite, Sonora, Mexico. They are the simplest type of vertical burrow and, though common in the Cambrian, are extremely rare in the Precambrian. Scale bar in centimeters.

brian in age. Steeply inclined burrows several centimeters deep called *Skolithos declinatus* are known from the late Precambrian of the Soviet Union (Sokolov and Ivanovskii 1985). But deep trace fossils such as these are exceptional in Precambrian sediments. Other late Precambrian tracemakers were able to shallowly penetrate sediments. *Neonereites* threaded a sine-wave-shaped path through the sediment. This tracemaker pratically swam through the sediment, something like a porpoise swimming through surface waters. A *Neonereites* fossil usually looks like a train of holes running across the bedding plane surface of a rock sample, where fracturing of the rock along a horizontal bedding plane surface has cut through the vertically sloping portions of the trace fossil.

Precambrian deposit feeding burrows are quite shallow, most being restricted to the upper centimeter of the sediment surface. The shallow depth of Precambrian deposit feeding burrows, like the near absence of Precambrian vertical burrows, is puzzling. With the beginning of the Cambrian, the depth of burrowing for both deposit feeding burrowers and dwelling burrow formers took a leap downward. *Skolithos* specimens 10 cm in length and longer have been reported from Early Cambrian rocks in Scandinavia and elsewhere. The increased depth of Cambrian deposit feeding is exemplified by deep, ploughing trace fossils from northern Mexico (figure 3.6). Traces such as these, in which an animal greater than one centimeter in diameter is able to muscle its way through the sediment, are unknown before the very end of the Precambrian.

Along with the increased depth of burrowing, there was an astonishing increase in the diversity of trace fossils at the beginning of the Cambrian. The number of different types of trace fossils soared at this time, and their abundance in any particular sedimentary environment also went up. *Skolithos* is so abundant in parts of the Early Cambrian Kalmarsund Sandstone of Sweden that it could be called "organ-pipe" rock. Arthropod trace fossils, some made by the earliest trilobites, became abundant for the first time. *Cruziana*, a trace fossil made by a deposit-feeding trilobite or trilobite-like organism as it "cruises" through the sediment, was commonly formed in the earliest Cambrian (figure 3.7). Deep furrowed traces appeared in abundance on the surface of sandstone deposits. Some of these fossils meander in complex, geometrically regular patterns virtually unknown in Precambrian traces. *Skolithos* is by no means the only new type of dwelling burrow. *Diplocraterion* appeared as a U-shaped burrow with two vertical arms, and other types of dwelling burrows

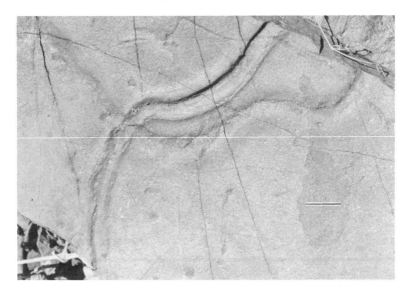

FIGURE 3.6. Horizontal trace fossil made by an animal ploughing at or just under the sediment surface at the sea floor. From the Lower Cambrian Puerto Blanco Formation of Sonora, Mexico. Scale bar = 2 cm.

FIGURE 3.7. *Cruziana*, a distinctive trace fossil with scratch marks made by the limbs of an arthropod such as a trilobite. Scale bar = 1 cm.

appeared, including one that flared upwards like an inverted cone. Some burrows such as *Phycodes* (figure 3.8) had fan-shaped arrays of deposit-feeding probe tracks radiating from a central dwelling tube. The tracemaker of *Phycodes* was based in the central tube and mined the sediment immediately nearby. This type of activity also forms star- or flower-shaped trace fossils (figure 3.9). Even the tubular deposit feeding burrows displayed some fancy new forms (Crimes 1987). *Treptichnus*, the "feather-stitch" trace fossil, consists of linked, alternating segment deposit feeding tubular burrows. The maker of *Gyrolithes* spiralled downward into the sediment like a corkscrew.

Even shallow, sediment surface burrows in the Cambrian show a marked change in character over their Precambrian predecessors. In the late 1840s and early 1850s, the oldest fossil known was *Oldhamia antiqua*, a delicate fan-shaped fossil discovered by Thomas Oldham in the Lower Cambrian sediments of Bray Head, Ireland (Secord 1986). These radiating or fan-shaped trace fossils imply a fairly complex "advance and retreat" type of deposit feeding behavior, and are known only from the Cambrian (Hofmann and Cecile 1981).

Net-shaped trace fossils or graphoglyptids (figure 3.10) are distinctive because of their unusually regular geometric pattern. The first graphoglyptid trace fossils appear in the earliest Cambrian. Graphoglyptid traces get their name from their resemblance to writing carved in stone, and indeed they can look like written messages from

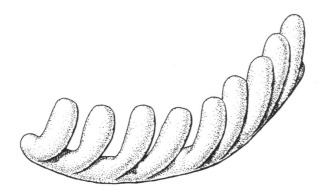

FIGURE 3.8. *Phycodes pedum*, a trace fossil formed by an animal digging through the sediment in search of buried morsels. The tracemaker made successive probes upward through the sediment in its attempts to find food. Width, including side branches, 4 cm. (From M. McMenamin 1989)

the history of past life. A typical graphoglyptid trace consists of a hexagonal network of cylindrical burrows a few millimeters below the sediment surface. This network connects to the open water by a series of tubes that rise upwards from the points where three of the horizontal canals meet to form one of the corners of a hexagon. Crimes and Crossley (1980) have shown that the graphoglyptid burrow geometry is engineered to make currents flow through the buried tunnel network. Flow through the burrow network is passive; in other words, it results from open water currents passing over the vertical risers, or pipelike tubes, in the graphoglyptid system. This aquatic ventilation of the buried tunnel system, a simple consequence of the burrow geometry, is of great use to the maker of the burrow. Unfortunately, the actual feeding strategy of the graphoglyptid tracemaker is unknown, even though modern graphoglyptid traces (identified as hexagonal patterns of holes, the terminations of the vertical tubes) have been spotted by cameras surveying the deep-sea floor (Weisburd 1986). A recent oceanographic expedition successfully collected an entire graphoglyptid burrow system in a sediment

FIGURE 3.9. A star-shaped trace fossil is visible to the right of the lens cap. From the Lower Cambrian Puerto Blanco Formation of Sonora, Mexico. Scale bar = 5 cm.

grab sample, but the sample was inadvertently washed down a drain by a technician before the tracemaker could be identified!

One hypothesis for the food-collecting strategy of the graphoglyptid tracemakers is filter feeding. The flow through the tunnel network functions anytime that there is a bottom current on the seafloor above, and a filter feeder would be supplied with suspended goodies without having to expend energy pumping water through filtering devices. Once the tunnels were built, passive flow from sea floor currents would provide all the energy necessary for filter feeding.

An alternative feeding strategy hypothesis has been proposed by Seilacher (1977), who suggests that the tracemaker's burrow system has been designed to cultivate a moneran "crop" on the interior burrow walls. Monerans are capable of breaking down resistant organic substances that metazoans cannot digest. Graphoglyptid animals may be using their burrow walls in much the same way that cows use microbes in their complex digestive system to break down indigestible plant substances such as cellulose (cows can digest grass

FIGURE 3.10. The graphoglyptid trace fossil *Protopaleodictyon*, from the Lower Cambrian of western Canada. Burrow geometry in many graphoglyptids is such that water flows continuously through the interconnected passages. (Photo courtesy J. Magwood)

and wood). Seilacher (1977, 1986) calls this the "mushroom farmer" hypothesis—the graphoglyptid animal uses its bacterial garden to gain access to otherwise indigestible nutrients in the sea floor sediments. Such organic compounds would include sporopollenin, a very resistant substance that composes acritarchs (unicellular fossils considered in chapter 8). The passive ventilation of the graphoglyptid burrow would be an asset to "mushroom farmers" as well as filter feeders, because a steady supply of fresh water could encourage the growth of the moneran "mushrooms." For food, the "mushroom farmer" need only graze the walls of its burrow. Other Early Cambrian trace fossils, like the downward spiralling *Gyrolithes*, may have also been designed for "mushroom farming" activity.

As the above examples demonstrate, something outstanding happened to the abilities of trace-fossil makers across the Precambrian-Cambrian boundary. Animals discovered a large number of ways to effectively use the sediment as a food resource, and also began to move deeper into the substrate for deposit feeding and homebuilding. Crimes (1987) shows that the diversity of trace fossils in shallow marine water (less than a few hundred meters deep) at the onset of the Cambrian is greater than at any time before, and, perhaps unexpectedly, greater than at any time since. This pattern is not paralleled by the shelly fossils of the earliest Cambrian; Cambrian shelly fossils are low in diversity compared to later times. Trace fossils such as graphoglyptids are characteristic of deep-ocean sediments in postCambrian rocks, but they are found only in shallow water settings in the Lower Cambrian.

It is easy to imagine why any given trace-fossil-type or specific feeding behavior might be restricted to a particular marine environment. *Skolithos* today is primarily found in nearshore, sandy environments where wave and tidal current energy is constantly disrupting or shifting the substrate. A deep vertical burrow (or the ability to form one quickly) is an asset to organisms living in shifting sandy substrates. Seilacher (1977) has argued persuasively that geometrically regular traces, such as those with tight horizontal meander loops (something like the loops of freon tubing on a refrigerator), show a behavior best suited to deep-sea deposit feeding where organic matter is scarce relative to that in sediments forming closer to the continents. Deep-sea deposit feeders avoid mining the same patch of sediment twice by never crossing their own tracks.

Less easy to understand is why so many types of burrowers, now restricted to radically different environments, were living together in shallow water at the Cambrian boundary. Why were all these different burrowing behaviors and types of feeding abruptly thrown together in the same environmental setting? This diversity is equivalent to finding polar bears, bighorn sheep, and elephants in the same national park. Something very curious caused animals to explore ways of living (or ecological niches) that had never been tested before. Was it some change in the environment that allowed animals to do this for the first time? Did animal populations suddenly get large enough to diversify in this way? Is this a record of a metazoan "big bang," when animals capable of doing these sorts of things evolved for the first time, and happened to find themselves in shallow marine water? Or did environments appropriate for this type of feeding behavior become available for the first time, thereby allowing animals to live in this way? These questions are all merely variations on the central mystery of the emergence of animals that we posed in the first chapter.

Small Shelly Fossils

A S DISCUSSED in the first chapter, the appearance of abundant shelly fossils has long been recognized as a puzzling discontinuity in the geologic record. What are these earliest shelly fossils? Were all of them formed by animals? This section will provide an introduction to some of the earliest shelly fossils. Although by no means exhaustive, the thumbnail sketches below will introduce all the major shelly animal groups that appear in the Early Cambrian. The first and last appearances of many members of these groups have been used by biostratigraphers to sequence the events of the Precambrian-Cambrian transition. Figure 4.1 shows the order of stratigraphic appearance of some of the Precambrian and Cambrian shelly fossils discussed here.

FIGURE 4.1. Stratigraphic ranges of some important Precambrian and Cambrian fossil animals. (After M. McMenamin 1987a)

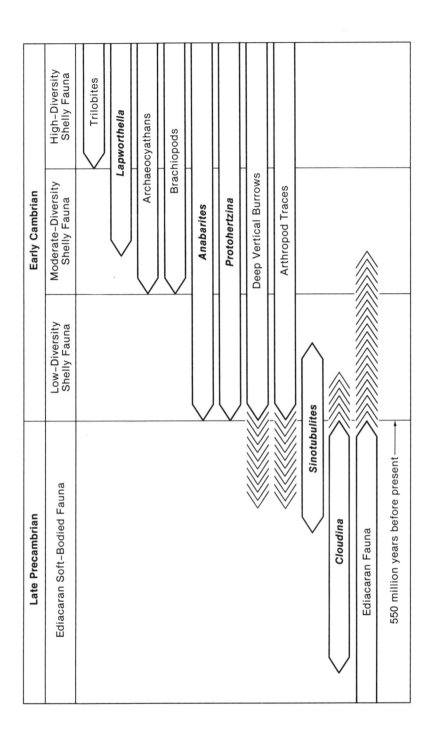

CLOUDINA

THE PRESENCE of Precambrian shelly fossils is, of course, dependent on where one places the boundary between the Precambrian and Cambrian in the global sedimentary record. Most geologists and paleontologists agree that the Ediacaran fauna should be considered Precambrian and that "typical" Cambrian fossils such as trilobites and brachiopods should be considered Cambrian. Using the presence of Ediacaran fossils as indicating Precambrian, there is only one genus of shelly fossils that seems to be unequivocally Precambrian in age. This is *Cloudina* (figure 4.2), named for Preston Cloud (Germs 1972). *Cloudina* was first described from the Nama Group of Namibia, the same sedimentary sequence that produced the first finds of Ediacaran soft-bodied fossils just after the turn of the century. *Cloudina* may prove useful for identifying Precambrian rocks in other regions, as this fossil is distinctive and appears to occur in widely separated regions of the southern continents that were once part of the supercontinent Gondwana (to be discussed further in chapter 6).

FIGURE 4.2. The earliest known calcium carbonate shelly fossil, *Cloudina*. It is from the Precambrian Nama Group of Namibia. Length of specimen 2 cm.

Zaine and Fairchild (1987) have recently illustrated a specimen of
Cloudina from Brazil that shows the distinctive "cone-in-cone" wall
structure of this creature's shell.

Cloudina is a tubular fossil up to a few centimeters in length. Its
shell was originally calcium carbonate in composition. Calcium carbonate shell comes in two varieties or mineral types, calcite and
aragonite. Calcite is a geologically common mineral and is the major
constituent of limestone. Shelly organisms commonly secrete calcite to construct shells. Aragonite, chemically the same as calcite,
differs radically in its crystal structure, and is less common than
calcite in ancient sediments, primarily because it is less chemically
stable. Buried aragonite, when subjected to the changes in acidity
and pressure that occur when rocks become cemented, may recrys-
tallize as calcite or dissolve altogether and return to water as carbonate ions. Aragonite is also commonly used by marine animals to
fabricate shells. Some mollusks, such as the edible mussel (*Mytilus*),
form a shell that has alternate layers of calcite and aragonite.

Cloudina fossils are found in platy limestone beds within the
Nama Group, sandwiched between quartzite (a tightly cemented
variety of sandstone) beds that have the impressions of soft-bodied
fossils such as *Pteridinium*. *Cloudina* tubes are often preserved bent
or kinked, suggesting that in life they may have been slightly flexible. Lowenstam (1980) feels that they were only partly mineralized,
and that they were not rigid, unflexing tubular skeletons like some
of the Cambrian shelly fossils that were to follow. *Cloudina* can be
seen easily with the naked eye (some tubes are four or more centimeters long). The best way to study the structure of fossils like
Cloudina is with thin sections, that is, with thin slices of fossiliferous rock ground flat on a lapidary wheel and mounted to glass slides
so that light passes through. Thin sections allow study of shell
microstructure with the aid of a transmitted light microscope. Shell
structure in *Cloudina* consists of tube wall layers arranged in cone-
in-cone fashion, like stacks of paper cups with their bottoms torn
out. This wall construction led Glaessner (1976) to conclude that
Cloudina is the fossil of a tube-dwelling annelid worm.

ANABARITIDS

ONE OF the oldest groups of Cambrian shelly fossils is the anabaritid
group (figure 4.3). Depending on where the boundary between the

Cambrian and the Precambrian is ultimately placed (it has not yet been formally defined), some anabaritids may prove to be latest Precambrian. Anabaritids were first discovered earlier in this century by a Swedish paleontologist who noted their distinctive "cloverleaf" cross sections in a sliced or slabbed piece of Cambrian limestone (S. Bengtson, personal communication). The genus *Anabarites* was first named and formally described in 1969 by the Soviet paleontologist V. V. Missarzhevskii in a book on the earliest Cambrian fossils of the Siberian platform (Rozanov et al. 1969). This volume has proven to be so important for subsequent study of Early Cambrian paleontology that many researchers refer to it as the "brown bible" (in reference to the brown cover of the original Soviet edition; an English translation, Rozanov et al. 1981, now exists, also with a brown cover).

Anabaritids are typical earliest Cambrian fossils. They are small —usually less that a few millimeters in greatest dimension. They are commonly found as phosphatic internal molds. An internal mold forms when the hollow interior of a fossil is filled by sediment (either as mud or silt particles, or by chemical precipitation). Many earliest Cambrian fossils are preserved as phosphate internal molds (the phosphate apparently formed as a chemical precipitate). These internal molds can be isolated from enclosing limestone sediment by dissolving the limestone with acetic acid (a relatively weak acid). Acetic acid also dissolves apatite, but so slowly that the phosphatized fossils can usually be recovered before they have been attacked or etched by the acid to any great extent. The fossils remain, plus sand grains and other minerals insoluble in the acid, in a residue

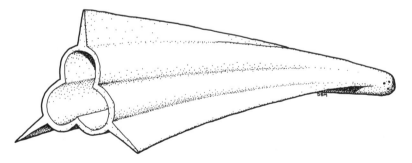

FIGURE 4.3. Reconstruction of *Anabarites,* an early Cambrian or late Precambrian shelly fossil with triradial symmetry. The function of the vanelike "stringers" present in some species is unknown. Length of shell 5 mm.

that is left after the spent acid is washed away. Great numbers of fossils can be obtained in this way. Fossils recovered in this way are generally referred to as "small shelly fossils." This is not a particularly informative designation (and sometimes gets perverted to "small silly fossils" or "small smelly fossils"), but is nevertheless appropriate because many of these fossils have little in common save their diminutive size.

Internal molds do not give much information about the original shell structure of anabaritids. Thin section study of actual shell material has shown that anabaritids had a thin, delicate aragonitic shelly wall. In some rare specimens that have preserved shell wall, the wall contains fine growth rings, and comes to a blunt termination at the tip or apex. In cross section, the shell is circular near its apex, and then develops three furrows as growth continues and the shell cone expands. In most specimens, these furrows deepen, giving the shell its distinctive "clover-leaf" cross section at the aperture or open end of the shell. In a few species, each lobe on the flank of the shell has a "stringer," or flat vane-like flange, extending from the outermost edge of the lobe (figure 4.3). The function of these stringers is unknown, but they may have served to strengthen a shell that in life was lying prostrate on the sea floor, or kept it from sinking into the substrate.

The nature of the anabaritid tube dweller is unknown. As with *Cloudina,* an elongate organism such as an annelid worm could have fit into this type of tube. There is some evidence, however, that the anabaritid animal was not an annelid worm. Annelids are bilaterally symmetric. Most of their appendages come in pairs, and there is a plane of symmetry running through the length of the body. In other words, the two halves of the body are (in most respects) mirror images of each other. Anabaritids may have been triradially symmetric in their soft parts as well as in their shell. That is, their body may have been divisible into three equal sectors, much like a cloverleaf. Some superbly preserved Australian anabaritid fossils have been studied by Simon Conway Morris (1987a). These fossils are not internal molds, but silicified reproductions of the thin, original shell wall. The calcium carbonate shell material was replaced during fossilization by silica (silicon dioxide, the formula for the mineral quartz), resulting in a complete and finely detailed specimen. Silicified fossils are easily removed from limestones, because the quartz is unaffected by acids that eat calcium carbonate.

Conway Morris (1987a) has found anabaritid specimens with three

51

tiny holes, symmetrically placed at their apex end, and he suggests that this may indicate that anabaritids originally had three-fold symmetry because they express the triradiate pattern so early in their life cycle. The triradiate shape of anabaritid aperture had earlier led the Soviet paleontologist M. A. Fedonkin (1981) to suggest an affinity between anabaritids and certain triradiate members of the Ediacaran fauna, such as *Skinnera* and *Tribrachidium*. Triradiate symmetry is very rare in modern animals, and if Fedonkin's (1981) speculation proves to be correct, it will provide an important link between the Ediacaran and the Cambrian fossil faunas.

OTHER SMALL SHELLY FOSSILS

IN ADDITION to anabaritids, numerous other conoidal fossils are known from earliest Cambrian limestone. One small fossil, *Tiksitheca*, can only be distinguished from anabaritids by the lack of the three furrows. A group related to anabaritids has six symmetric furrows instead of just three. Not all of the early shelly fossils are cone-shaped with pointed ends. Some are cylindrical to very weakly tapering tubes. *Coleoloides* is such a tube with a calcium carbonate (originally aragonitic) shell. These shells may be straight to strongly curved, and sometimes bear an attractive spiral ornament (figure 4.4). In Lower Cambrian limestones of southeastern Newfoundland, *Coleoloides* tubes are fossilized clustered together, all oriented vertically, presumably in their original life position. The Newfoundland fossils are very curious, and numerous horizons in the limestones display thousands of these fossils clustered together, looking like straws (about a millimeter in diameter and up to several centimeters long) stuck in the mud.

Another tubular fossil is one of the earliest known to have originally had a calcium phosphatic composition. This is *Hyolithellus*, an elongate tube indistinguishable from some specimens of *Coleoloides* except for its phosphatic composition. *Hyolithellus* has rings or annulations in most species (figure 4.5). Although the tube may taper somewhat, an apex has never been found.

Protohertzina is another one of the first shelly fossils with a phosphatic composition (figure 4.6). This tusk-shaped, solid phosphatic (not hollow) fossil is found with anabaritids in limestone beds that contain the earliest Cambrian-type shelly fossils in many stra-

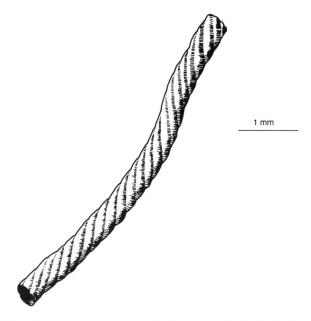

FIGURE 4.4. Some species of the early Cambrian shelly fossil, *Coleoloides,* bear a distinctive spiral ornament. These calcitic tubes are often found in clusters. Scale bar = 1 mm. (After Matthews and Missarzhevsky 1975)

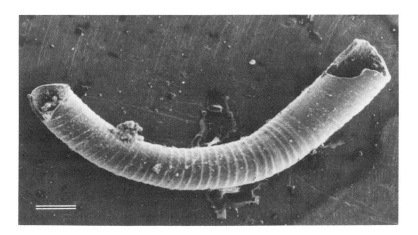

FIGURE 4.5. *Hyolithellus,* a widely distributed Cambrian small shelly fossil with a shell composed of apatite, a phosphatic mineral. From the Puerto Blanco Formation of Sonora, Mexico. Scale bar = 0.2 mm.

53

tigraphic sections throughout the world. *Protohertzina* belongs to the protoconodonts, a group of spine-shaped phosphatic fossils. Interestingly, Szaniawski (1982) has shown that there are strong similarities between the microstructures of some protoconodonts and the grasping spines of modern chetognaths (pronounced KEET-o-naths) or arrowworms. Chetognaths are tiny but voracious marine predators that have sharp grasping spines near their mouths for seizing prey. Fossil chetognaths first appear in the Carboniferous. Assuming that protoconodonts and chetognaths are indeed related, *Protohertzina* is the earliest convincing evidence for a fossil metazoan predator.

Many of the conical or pyramidal fossils from the Early Cambrian with shells composed of calcium carbonate have been called hyoliths. Hyoliths (figure 4.7) are a problematic group of fossil shells that went extinct in the Permian. The conical hyolith shell had a "trap door" or operculum, and—in well-preserved specimens—lateral "arms," or helens, project outward from the junction of the aperture and the operculum. They have been placed in the mollusk phylum (clams, snails, and squids) by some researchers, but have been put in their own phylum by others (Pojeta 1987). Some hyoliths such as *Burithes cuneatus* from the Tommotian Stage of Siberia (Rozanov et al. 1981:157) are distinguished by the "lip" or ligula projecting from the lower part of the aperture of the conoidal shell. Other hyoliths, called orthothecids, lack the ligula, which has led Yochelson (1979) to question whether or not these are truly hyoliths. Yochelson (1979) notes that a conical shell is one of the simplest shell plans, and there is every reason to believe that non-hyolith organisms could "invent" this shape of shell with an operculum.

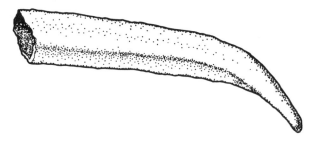

FIGURE 4.6. The interior of *Protohertzina* is solid rather than hollow. This phosphatic Lower Cambrian shelly fossil is from the earliest known metazoan predator. Length of fossil approximately 1.5 mm.

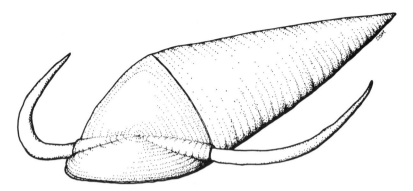

FIGURE 4.7. Reconstruction of a hyolith. The tusklike appendages are called helens, and the cap on the wide end of the conical shell is called an operculum. Some researchers place hyoliths in phylum Mollusca. Length of shell 1 cm.

SCLERITOMES

MOST OF the fossils just discussed are fairly simple and nondescript, at least compared to the broad spectrum of shell groups among modern marine organisms. This simple shell form makes classification difficult. Just as numerous types of animals can make a simple tubular burrow, many types of animals can create a simple tubular or conical shell. But the classification problems with tubular and conical shells are not nearly as daunting as are those for a group of Early Cambrian fossils known as sclerites. Sclerites are individual mineralized skeletal elements that once formed part of a multi-element external skeleton for an animal that was originally much larger than any individual sclerite. A sclerite would be analogous to a single metal ring from a chain-mail coat. A fish scale is a modern example of a sclerite. The entire collection of sclerites from the body of the organism which formed the sclerites is called the scleritome (body of sclerites). All the scales from the skin of a fish would constitute the fish scleritome.

The individual spines of a sponge's skeleton could be considered sclerites, except that many of these spines visible at the surface of the sponge are deeply embedded in the body of the sponge, and the word "spicule" is preferable for skeletal elements of this sort. Sponge spicules support the soft tissue that forms the filter feeding chamber, used by the sponge to strain the seawater for microscopic food particles. Sponges with calcium carbonate, siliceous, and organic spicules all first appear in the Cambrian. The most ancient sponge spicules are known from near the beginning of the Cambrian from Siberia (Matthews and Missarzhevskii 1975). These six-rayed spicules look very much like a child's jacks, but are much smaller, being less than a millimeter in length.

Many earliest Cambrian shelly fossils are thought to represent fragments of ancient scleritomes. Some of these skeletal elements are conical, cap-, or cornucopia-shaped, and were first thought to be complete shells that housed an entire tiny animal in the sclerite cavity. This was an early interpretation for the shelly fossil *Lapworthella* (figure 4.8a). The "whole shell" interpretation is now known to be false, however, because in some *Lapworthella* and related sclerites the cavity or "living chamber" is completely eliminated (Bengtson 1970). Also, *Lapworthella* sclerites have been found which are fused together in a manner which would be impossible if each sclerite represented a single, free-living individual (figure 4.8b). The sclerite fusion is best interpreted as accidental joining of adjacent sclerites that grew as part of the same scleritome (Landing 1984). *Lapworthella* had an originally phosphatic composition, as did a number of other sclerite genera collectively known as tommotiids (named for the Tommotian Stage, a Lower Cambrian stratigraphic unit of the Siberian platform).

Tommotiid sclerites often come in two varieties that are now thought to have originally belonged to the same type of animal. *Tommotia* is a cap-shaped phosphatic tommotiid sclerite. In many acid-residue samples that produce *Tommotia*, there also occurs a saddle-shaped or sellate sclerite that has been called *Camenella*. Although their external shape varies, the ornamentation on the two types of sclerites in any particular acid residue sample is very similar. This led Bengtson (1970) to conclude that *Tommotia* and *Camenella* were actually part of the same scleritome. Unfortunately, re-

FIGURE 4.8. A: *Lapworthella filigrana*, a Lower Cambrian sclerite fossil, from the Puerto Blanco Formation of Sonora, Mexico. Greatest dimension of sclerite 1.6 mm; B: *Lapworthella schodackensis*. Two ontogenetically fused sclerites of dissimilar sizes forming a composite sclerite. These sclerites are compelling evidence that *Lapworthella* sclerites were once part of a multi-element scleritome. Height of larger sclerite 0.47 mm. Same specimen as illustrated in figure 2M of Landing (1984). (Photograph courtesy Ed Landing)

construction of the original scleritome with all its pieces in place is not an easy task. Individual sclerites do not fit together like pieces of a jigsaw puzzle, and in almost every case not all of the pieces of any puzzle are available to the paleontologist. An acid residue with sclerites is like having one or two pieces from each of a dozen different jigsaw puzzles. Sclerite reconstructions have nevertheless been presented for several genera. Figure 4.9 shows a reconstruction for the genus *Lapworthella*. As interpreted here, the sclerites of *Lapworthella* studded the surface of a slug-like animal that was a centimeter or two in length.

In addition to the phosphatic sclerites, a number of calcium carbonate Cambrian shelly fossils are now interpreted as representing isolated sclerites. The sclerite *Chancelloria* is very commonly found in Early Cambrian acid residues. A natural association of *Chancelloria* sclerites on a bedding plane surface is shown in figure 4.10. Each sclerite is shaped something like the hub and spokes of a wagon wheel, except that the spokes often curve away from the hub instead of radiating straight and in the same plane. Six hollow spines radiate out from the central boss, and sometimes a seventh spine projects outward, perpendicular to the bases of the other six spines. *Chancelloria* spines were aragonitic in life, but they have a strong tendency to be phosphatized (replaced by phosphate) and are thus very commonly preserved. *Chancelloria* was originally described as a variety of sponge spicule, but the spine structure of *Chancelloria* has very little in common with any known sponge spicule. Bengtson and Missarzhevsky (1981) suggested that *Chancelloria* sclerites were ele-

FIGURE 4.9. Reconstruction of the *Lapworthella* animal and its scleritome. Length of animal 9 mm. (McMenamin 1987a; copyright © 1987 by Scientific America, Inc. All rights reserved)

ments of a scleritome, and this is almost certainly the case. Sometimes *Chancelloria* sclerites are associated with a three-pronged sclerite of the same construction that has been called *Allonia*. *Chancelloria* and *Allonia* sclerites were probably once part of the same scleritome.

Chancelloria may be the only known Cambrian mineralized sclerite for which there exists a scleritome preserved more or less intact. Walcott (1920) described organized clusters of *Chancelloria* spicules (figure 4.10) from the now famous Burgess Shale, a Middle Cambrian shale deposit which has yielded soft-bodied and shelly fossils of unequalled importance for the study of early animals (Whittington 1985). Some clusters of *Chancelloria* sclerites also contain specimens of *Allonia*, further supporting the suggestion that they once belonged to the same animal. The spiny coat of sclerites probably served a protective function for the presumably metazoan *Chancelloria* animal within.

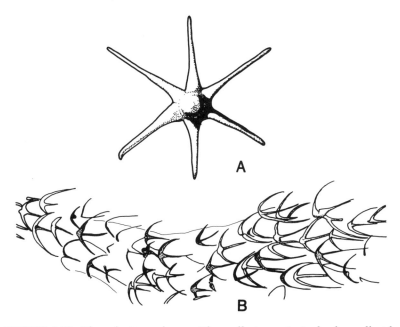

FIGURE 4.10. The scleritome-bearer *Chancelloria*. A: A single chancellorid sclerite, width from spine tip to spine tip about 6 mm; B: a sclerite map of a more-or-less intact *Chancelloria* scleritome from the Burgess Shale of British Columbia, Canada. Each anchor-shaped object in this phostograph is a flat-tened sclerite. Length of scleritome 27 mm. (After Walcott 1920)

Another intact scleritome, this time unmineralized, is known from the Burgess Shale. The scleritome of *Wiwaxia* has been studied in depth by Conway Morris (1985), and he used it as "template" for the reconstruction of the scleritome of the calcium carbonate mineralized sclerite *Halkieria*. The nonmineralized spines of *Wiwaxia* and the mineralized sclerites of *Halkieria* are roughly similar in shape, leading Bengtson and Conway Morris (1984) to propose a scleritome for *Halkieria* that is similar to a wiwaxiid scleritome. The wiwaxiid scleritome is a mosaic of tightly fitting, overlapping scales, with two rows of elongate scales projecting menacingly upward from the flat-lying sclerites. Cambrian scleritome bearers must have resembled tiny marine hedgehogs or porcupines.

ARCHAEOCYATHANS

PERHAPS THE most distinctive and yet puzzling of Cambrian shelly fossils are the archaeocyathans, also called archaeocyathids. Archaeocyathans have a cup-shaped, originally calcareous, skeleton with porous, double walls connected by partitions called septa (figure 4.11). The walls and septa are a meshwork of connecting rods

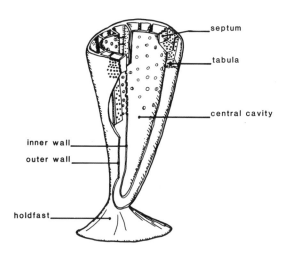

FIGURE 4.11. Cut away diagram showing the structure of a typical Lower Cambrian archaeocyathan. Diameter of cup 2 cm. (After Debrenne 1964)

and porous plates that are not at all like the spicular skeleton of a sponge. Not all archaeocyathans follow this basic pattern shown in figure 4.11. Some forms are flattened out to form saucer-like objects, and others branch and twist into bizarre, irregular shapes (Hill 1972). "Normal" archaeocyathans superficially resemble sponges and solitary corals, but their detailed structure is so unlike any known group of organisms that most scientists now place them in a separate phylum. It is not at all clear that archaeocyathans were animals, and in fact they may have been more closely allied to marine algae with calcium carbonate skeletons (Nitecki and Debrenne 1979). The feeding strategy of archaeocyathans is not known with certainty, but Balsam and Volgel (1973) have suggested that archaeocyathans were passive filter feeders. This means that they strained water for food particles but didn't actively pump water through their bodies.

Balsam and Volgel (1973) constructed a stainless steel model of an archaeocyathan. When the model was placed into a current of water with a stream of dye as a marker, the water flow (as indicated by the dye) went through the pores into the archaeocyathan and flowed out and upward through the hole made by the top of the inner wall. An alternative suggestion was first made by Richard Cowen, who suggests that archaeocyathans may have lived in association with photosymbiotic monerans (Cowen 1986). Rowland (1988) carries this idea further, and provocatively suggests that archaeocyathans are skeletonized relicts from the Ediacaran fauna!

Archaeocyathans are crucial for the study of early animals for two reasons. First, they are one of the few groups of Early Cambrian organisms that undergo rapid enough evolution in skeletal form to be useful for high resolution biostratigraphy. As a result, they have been used as the basis for subdividing the Lower Cambrian into stages on the Siberian platform. Some paleontologists have suggested that the Soviet biostratigraphic subdivision scheme should be adopted as a worldwide standard for naming subunits of the Lower Cambrian, but this view is falling out of favor because strata in other parts of the world may provide a more complete stratigraphic record of the Lower Cambrian. Nevertheless, archaeocyathans are one of the most promising types of fossils for global correlation of the upper half of the Lower Cambrian. The oldest archaeocyathans occur at or very near the base of the Cambrian as currently accepted by many paleontologists. Unfortunately, these very early archaeocyathans are known only from the Siberian platform.

Archaeocyathans are also partly responsible for the earliest skeletal reefs. Skeletal reefs are, in essence, piles of calcium carbonate shell material (sometimes the shells are cemented together). Such shell accumulation can interfere with nearshore ocean circulation and prevent the full force of marine waves from hitting the shoreline. Reefs can thus create sheltered areas called lagoons that occur between the reef and the shoreline. Archaeocyathans and several types of Cambrian calcium carbonate branching algae grew together to form early reefs (Rowland 1984). The reef environment and associated lagoons were important habitats for the evolution of early animals. The photosymbiosis hypothesis (Cowen 1986) for archaeocyathans is consistent with their reef-forming habits. Many important reef-forming marine organisms live (or, for extinct forms, are thought to have lived) in association with photosymbiotic monerans.

TRILOBITES

THE BEST known and most typical Cambrian fossils are trilobites. The familar Cambrian species *Elrathia kingii* (figure 4.12) has been quarried in great quantities from the Wheeler Shale of western Utah and can be found all over the United States in rock shops and museum stores, often fashioned into bolo ties and earrings. The Ute Indians of Utah made specimens of *Elrathia* into amulets, calling them by the Ute name *timpe khanitza pachavee,* which roughly translates to "little water bug that likes to live in a stone house" (Robison 1987).

The popularity of trilobites may be due to the fact that they bear the most ancient visual systems known—they are the first fossil animals with eyes. It is perhaps easier for us visually oriented humans to relate to fossil organisms that appear to stare back at us from their silty tomb.

In addition to eyes, trilobites have a hard dorsal external skeleton (the carapace), numerous jointed legs and one pair of antennae. A trilobite's carapace is divided into the cephalon (head), the thorax, and the pygidium (figure 4.12). Because of the jointed appendages, trilobites are frequently placed in the arthropod phylum. This phylum includes crabs, insects, horseshoe crabs, spiders, and other animals with external skeletons and rigid, jointed tubes for legs. Some paleontologists place trilobites in their own phylum because of their differences with respect to all other arthropods. By now, it must seem that the biological category (or taxon) "phylum" is a rather flexible concept, and indeed it is. "Phylum" is a difficult concept to define (Bengtson 1986), but a working definition might be as follows —"a phylum is a major group of organisms that are more closely related and more similar to one another than they are to anything else." This is a slippery definition, because it begs the question of what Bengtson means by "major." Although subjective criteria play a big role in determining what is and what is not a phylum, phyla are generally thought of as being just a step below the five grand divisions (kingdoms) of the earth's biota discussed in the first chapter. Identifying the phylum-level affinities of ancient fossil organisms can be difficult because crucial clues such as reproductive behavior, larval form, and soft body morphology are missing.

Trilobites had carapaces mineralized with calcite, in contrast to many of their shelly contemporaries that had aragonitic or phosphatic shells. Like most arthropods and snakes, trilobites had to shed

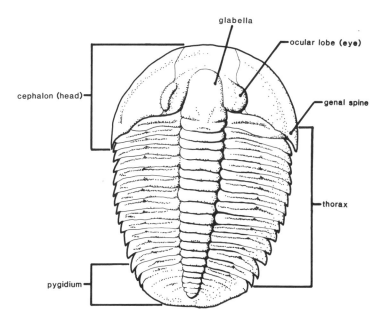

FIGURE 4.12. *Elrathia kingi*, a common Middle Cambrian tilobite, labeled to show the cephalon, thorax, genal spine, pygidium, and ocular lobes (eyes). Length of trilobite approximately 4.5 cm.

their skins occasionally to provide room for further growth. Trilobite fossils are thus of two varieties—corpses (the dead animal itself) and molts (cast-off carapaces). Lower Cambrian trilobites are perhaps best characterized by describing their inferred feeding behavior. They were, in essence, deposit feeding factories, consisting of a mineralized but flexible carapace above and many pairs of legs below. The legs are the more robust half of a double appendage—the top half of the appendage had delicate gills. The mouth is underneath the "head" end of the carapace, and is directed backwards. When a trilobite feeds, its legs stir up the sediment underneath the carapace. Lighter, organic-rich items in the sediment slurry are moved forward and sucked into the trilobite's mouth (Seilacher 1985). The elongate bulbous structure (called a glabella) between the trilobite's eyes that looks as if it ought to be the creature's braincase is actually its stomach (Cisne 1975).

In most parts of the world with Precambrian-Cambrian stratigraphic sections, the first trilobites are preceeded by other types of shelly fossils. For example, archaeocyathans of the Tommotian Stage on the Siberian platform occur before trilobites. Trilobites do not appear until the next stage of the Siberian Lower Cambrian, the Atdabanian Stage. In northern Mexico, a diverse suite of calcium carbonate and phosphatic small shelly fossils occurs stratigraphically below the oldest trilobites. Trilobites appear before small shelly fossils and archaeocyathans in only a few sections. One of these is the Moroccan section in the Anti-Atlas Mountains of northern Africa (Geyer 1988). This section may contain the oldest known trilobites.

For most Precambrian-Cambrian sections, there is a problem concerning the first appearance of trilobites. Trace fossils such as *Cruziana* (figure 3.7) presumed to have been made by the deposit feeding activities of trilobites often occur well below the first trilobite body fossils (Crimes 1987; Jenkins 1988). Alpert (1977) was first to analyze this problem, and he suggested that the first appearance of "trilobitoid" trace fossils be used to demark the Precambrian-Cambrian boundary. Where are the body fossils of these multi-legged trilobite-like organisms which were probably the ancestors of true trilobites? Perhaps their carapaces were not sufficiently well mineralized to preserve easily as fossils. Paleontology would be well served by a discovery of a Precambrian soft-bodied fauna that contained body fossils of the trilobite ancestors.

Trilobites are so useful for biostratigraphic subdivision of the

Cambrian that it is useful to describe the main trilobite varieties. Two types of trilobites appear in the Early Cambrian. These are the polymerids and the agnostids. Polymerids are "normal," full-size trilobites; agnostids are tiny forms probably specialized for a free-floating (planktonic) existence. The first trilobites to appear are polymerids.

Olenellid and redlichiid trilobites are among the earliest polymerids. They are characterized by a large semicircular cephalon and a tiny pygidium. *Judomia* and *Nevadia* are early olenellid trilobites known from Siberia and North America (figures 4.13 and 4.14). *Eoredlichia* is an equally ancient redlichiid trilobite known from the other side of the world in China. Some olenellids, such as *Laudonia*, are distinctive because of their long spines (figure 4.15). Ebenezer Emmon's *Elliptocephala* (figure 1.2) is another example of an Early Cambrian olenellid trilobite.

Ptychopariid trilobites are an important group of polymerids that first appear in the Early Cambrian; the Middle Cambrian species *Elrathia kingii* (figure 4.12) is the best known ptychopariid. Ptychopariids generally have a large thorax and smallish (but not tiny) pygidium.

ECHINODERMS

ECHINODERMS ARE a phylum of spiny-skinned, exclusively marine animals that first appear in the Early Cambrian. Modern examples include starfish, sea urchins, crinoids ("sea lillies"), and sea cucumbers. All modern examples have five-fold or pentameral radial symmetry (like a five-pointed star), but this is not the case for many fossil echinoderm groups. Echinoderms are most easily recognized as fossils by their distinctive skeleton composed of plates. Echinoderm plates qualify as sclerites, but individual plates in some echinoderms (especially specialized, modern sand dollars) are fused together to such an extent that it is difficult or impossible to make out the outlines of the individual plates. In these cases it is better to refer to the entire echinoderm skeleton as a "test" rather than as a scleritome. Echinoderm skeletons and skeletal plates are always calcareous. Each plate is a single porous crystal of calcite. These porous plates are unmistakable when viewed in thin section.

Three very different examples of echinoderms are known from

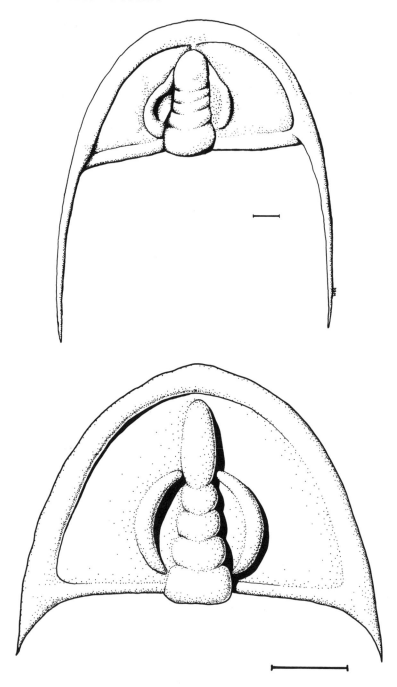

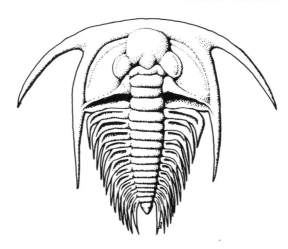

FIGURE 4.15. *Laudonia*, a Lower Cambrian trilobite with conspicuous metagenal spines. Greatest width 3 cm. (From M. McMenamin 1989)

the Early Cambrian: eocrinoids, edrioasteroids, and helicoplacoids. All three of these echinoderms are considered by Sprinkle and Kier (1987) to have been suspension feeders or filter feeders. The oldest echinoderm fossils known come from the Montenegro Member of the Campito Formation in the White-Inyo Mountains of eastern California (Durham 1971). These fossils are disarticulated plates that probably belonged to helicoplacoids (P. W. Signor, personal communication). Fossil eocrinoid plates occur slightly higher in the White-Inyo Mountains' stratigraphic succession. Eocrinoids (figure 4.16) are early stalked echinoderms, in which the main body of the animal is held above the sea floor by a supporting stalk. Despite their name, eocrinoids were not directly ancestral to the familiar crinoids which became extremely abundant 200 million years later (Sprinkle and Kier 1987). The stalk of the eocrinoid is composed of many small plates. This stalk supports the theca or main body of the eocrinoid. The theca is a sac-like or box-like body made of sutured or imbricate

FIGURE 4.13. Cephalon of the Lower Cambrian olenellid trilobite (*opposite above*), *Judomia*. Note its large eyes and long genal spines. Width of specimen (spine tip to spine tip) 3.6 cm. (From M. McMenamin 1988)

FIGURE 4.14. Cephalon of *Nevadia ovalis*, (*opposite below*), another large-eyed trilobite from the Lower Cambrian. Scale bar = 5 mm. (After M. McMenamin 1987b)

plates enclosing and protecting the main body. Thecal plates are larger than stalk plates and often have notches along their edges (figure 4.16), forming pores which ran into the interior of the theca. These sutural pores were used for respiration (Sprinkle and Kier 1987). Extending from the theca are eight or more arms (also composed of plates), which were used by the animal for filter feeding. In the eocrinoid *Gogia spiralis*, each arm is tightly spiralled (Durham 1971). The inner edge of each arm contains a food groove (or ambulacral groove). This food groove conducted trapped food particles to the eocrinoid's mouth, which was located at the top of the theca near the base of the arms.

Edrioasteroids also first appear during the Early Cambrian. Edrioasteroids have a biscuit-shaped, discoid, or globular theca that was cemented to the seafloor (or, occasionally, to another organism's shell). On the upper surface of the theca were five curving or straight food grooves. These food grooves make edrioasteroids appear similar

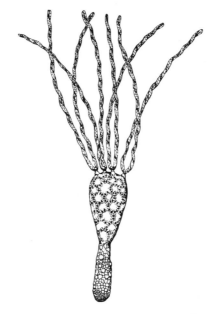

FIGURE 4.16. The Cambrian eocrinoid, *Gogia spiralis*. The name of this species refers to its unusual spiralled arms. An eocrinoid's stalk is make of many calcitic plates, while its calyx or "head" is made of beadlike plates. Despite the name, eocrinoids are probably not ancestral to true crinoids or "sea lillies." Length of specimen 5.3 cm. (From Robison 1965)

to *Tribrachidium* with its three radiating arms (Stanley 1976), but many paleontologists consider these similarities to be superficial. Gehling (1987) recently described, from the classic Ediacaran strata of Australia, a globular fossil with five rays on its surface that looks very suggestive of an edrioasteroid. This fossil, *Arkarua adami*, (figure 4.17) is presented by Gehling (1987) as the oldest known echinoderm, but its similarities to later echinoderms may prove superficial.

The most bizarre of the Early Cambrian echinoderms was *Helicoplacus*, a spiralled, spindle-shaped fossil restricted to Lower Cambrian sediments of western North America (figure 4.18). The whole test could be expanded and contracted; expansion apparently was accomplished by inflation from the inside. In one species, there are spines on some of the plates. *Helicoplacus* had one or more food grooves like the eocrinoids, except that instead of being on the arms, the grooves wound around the body of the animal, following the spiralled rows of plates. These grooves in *Helicoplacus* were presumably for filter-feeding, but this cannot have been a very efficient way to feed, with so little food groove surface area exposed. Indeed, there is controversy concerning the position of the mouth in *Helicoplacus*. Durham and Caster (1963) reconstructed the mouth position as being on the top of the pear-shaped or fusiform body. Derstler (1982) disagrees, arguing that the mouth was on the side of the spindle. Another possibility is that *Helicoplacus* had no mouth at all—perhaps it absorbed dissolved nutrients directly from sea water. Whatever their feeding strategy, helicoplacoids were a very short-lived group. They did not survive the Lower Cambrian.

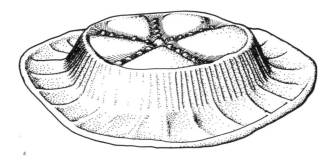

FIGURE 4.17. *Arkarua adami*, described as the oldest known echinoderm. Its affinities to true, skeleton-bearing echinoderms remain to be proven. Diameter of fossil 6 mm. (After Gehling 1987)

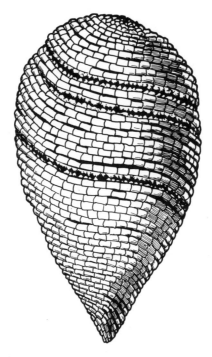

FIGURE 4.18. *Helicoplacus gilberti*, a Lower Cambrian echinoderm. Helicoplacoid food grooves wound in a spiral around the body of the animal parallel to the rows of plates, but the position of its mouth (where the food grooves joined together) is a matter of debate. Length of specimen 2.5 cm. (After Durham and Caster 1963)

BRACHIOPODS

BRACHIOPODS ARE important constituents of Early Cambrian shelly faunas. Brachiopods are bivalved, exclusively marine, organisms. They survive today but are not as common as they have been in the geologic past. All brachiopods past and present have been filter feeders, with the exception of a few that are thought to have had photosymbionts. The two valves of a brachiopod shell enclose the body of the brachiopod in much the same way that a clamshell encloses the body of a clam. Filter feeding is accomplished inside the shell by means of a lophophore, a coiled tentacle or pair of tentacles that trap food particles and move them to the mouth in a manner not unlike the food trapping and transporting method of the echinoderm food groove. The lophophore is sometimes held rigid by a delicate wirelike coil of skeletal material. This fragile apparatus is protected by the bivalved shell, and efficient filter feeding is possible by careful direction of the flow of water that enters through the commissure or opening between the two valves.

Brachiopods are divided into two groups that differ in the way the two halves of the shell are connected. Inarticulate brachiopods have a simple ligament joining the two shells. The ligament is composed of a tough, flexible, organic material. Articulate brachiopods, as their name suggests, characteristically have a skeletal articulation such as hinge teeth with complementary sockets connecting the two valves. Inarticulates are generally considered less "advanced" than the articulates (Clarkson 1979), and so it is not surprising that they appear first in the fossil record. The oldest brachiopods are known from the base of the Tommotian strata on the Siberian platform. Equally ancient brachiopods may occur in China and elsewhere. These earliest inarticulates (figure 4.19) are both calcium phosphate (phosphatic) and calcium carbonate in composition; phosphatic brachiopods are commonly found in Lower Cambrian acid residues. A very common early genus is *Lingulella*, a phosphatic form with a lozenge-shaped shell. *Kutorgina* is an early calcium carbonate inarticulate. Paterinids are phosphatic inarticultes that sometimes show fancy scalloping on the shell surface. Mickwitziids are another phosphatic group with abundant shell pores or punctae that will be discussed in later chapters.

The oldest articulate brachiopods are known from Siberia (Ushatinskaya 1986) and North America (Rowell 1977). These brachiopods have been tentatively assigned to the genus *Nisusia* (figure 4.20). The Soviet example has small bumps on the outer shell surface that were probably the bases of spines (Ushatinskaya 1986). Most of these Cambrian articulates and inarticulates probably spent their lives unmoving on the sea floor, straining the water for food. Most were also very small, averaging less than one centimeter in shell diameter. Mickwitziid brachiopods, another exclusively Cambrian group, were unusually large inarticulates; individuals attained sizes up to 3.7 cm in shell width.

MOLLUSKS

THE FINAL major group of Cambrian shelly fossils are the mollusks. Mollusks are the most familiar and successful phylum of shelled

71

invertebrate animals in modern seas, including clams (bivalves), snails (gastropods), squids and octopuses (cephalopods), chitons and a number of less important groups. Several major mollusk groups make their debut in the Early Cambrian, and most of the groups that appear subsequently can be thought of as descendants of the first Cambrian mollusks. The Early Cambrian mollusks are usually very small and all have calcium carbonate shells. Cambrian mollusks found in acid residues are primarily phosphatized internal molds or silicified shells. Since mollusks are such an important group of marine animals in the fossil record, a good deal of attention has been

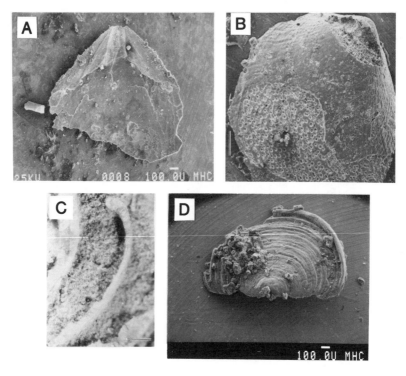

FIGURE 4.19. Several Lower Cambrian inarticulate brachiopods, from the Puerto Blanco Formation of Sonora, Mexico. A: interior of the valve of the phosphatic *Lingulella*; B: partially exfoliated valve of *Mickwitzia* (note abundant punctae); C: cross section through the calcium carbonate valve of *Kutorgina*; D: valve (exterior surface) of the phosphatic brachiopod *Paterina*. Scale bars: A: 0.1 mm, B: 0.25 mm, C: 1 mm, D: 0.1 mm.

focused on their beginnings. Some interpretations of these early fossils have generated heated controversy.

Three main mollusk groups appear together at the beginning of the Early Cambrian: gastropods, monoplacophorans, and rostroconchs. Possible Cambrian gastropods with the familiar coiled snail-shell aspect include *Aldanella* and *Pelagiella*. Some paleontologists have questioned whether or not these coiled shells actually represent snails, but in the absence of strong evidence to the contrary, this seems to be a reasonable assumption. Figure 4.21 shows a beautiful specimen of *Pelagiella* from the Lower Cambrian Buelna Formation of Sonora, Mexico. The delicate ornament of the shell surface has been preserved because of silicification of the thin, originally calcium carbonate, shell wall.

Monoplacophorans (figure 4.22) are cap-shaped shells distinguished by two rows of muscle scars on the interior of the shell. They were thought extinct until living specimens were dredged from the deep sea and described in the late 1950s. Monoplacophorans have had an unusual history of discovery. They are the only group of animals that has been: (a) described hypothetically before being discovered; (b) found as fossils before being found alive; and (c) dredged from the depths of the oceans before being collected from shallower marine waters (Pojeta et al. 1987). Monoplacophorans are rare and unimportant in modern seas; only two living genera *(Neopilina* and

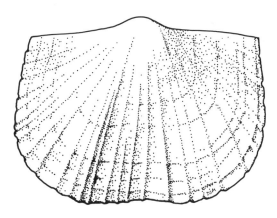

FIGURE 4.20. The Lower Cambrian articulate brachiopod *Nisusia sulcata*. The oldest articulated brachiopods probably belong to this genus, and some species may have borne spines. Width of valve 1.4 cm. (After Rowell and Caruso 1985)

Verma) exist today. Most of the earliest monoplacophorans were simple cap-shaped shells, with or without significant curvature (figure 4.22). One Middle Cambrian cap shell, interpreted by Pojeta et al. (1987) as a monoplacophoran, had a long snorkel projecting from the front of the shell (figure 4.23). The snorkel in *Yochelcionella* was probably an adaptation that improved circulation inside the shell.

Rostroconchs are a major, extinct, order of mollusks that first appeared in the earliest Cambrian. Rostroconchs have a shell that is shaped like a clam shell, except that instead of having an organic ligament connecting the two valves, the two halves of a rostroconch shell are fused together to form a single valve. Despite this fusion, larger rostroconchs look very much like clam fossils with valves still articulated, which partly explains why rostroconchs were not recognized as a major, distinct, group until the 1970s. Early rostroconchs had a plate, called the pegma, connecting the two halves of the shell (the pegma is not visible from the outside). The oldest rostroconch

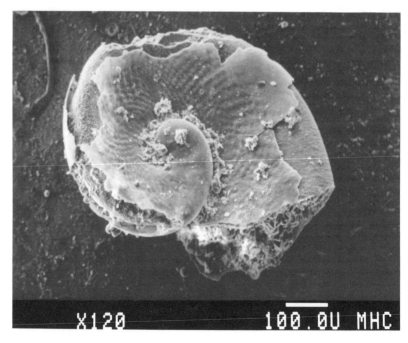

FIGURE 4.21. *Pelagiella*, a Lower Cambrian gastropod with original shell ornament preserved, from the Buelna Formation of Sonora, Mexico. Scale bar = 0.1 mm.

is *Heraultipegma* (figure 4.24). *Heraultipegma* has a worldwide distribution in Lower Cambrian rocks, being known from Australia, China, USSR, Europe, and possibly North America (Pojeta 1981).

Slightly after the first appearance of rostroconchs, the first true clams or bivalves appear. Clams probably had the same ancestor as the rostroconchs (Morris 1979). Instead of keeping the two valves fused as in rostroconchs, clams hinged the valves with articulating teeth and a tough, organic ligament. This evidently proved to be the more successful approach, since bivalve shells now litter the beaches all over the earth, whereas rostroconchs dwindled to extinction in the Permian. The oldest bivalves are the Lower Cambrian genera *Pojetaia* (figure 4.25) and *Fordilla*. Although their shells are only a few millimeters or less in greatest dimension, the presence of a ligament, clam-style muscle scars, and hinge teeth have been demonstrated in these genera (Pojeta 1981).

All of the known Early Cambrian mollusks had either a single

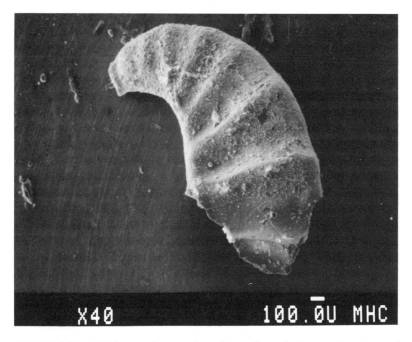

FIGURE 4.22. A high spired monoplacophoran from the Lower Cambrian of the Cassiar Mountains, Canada. Scale bar = 0.1 mm. (Specimen courtesy of S. Conway Morris)

shell or a pair of valves, and none of them are known to have had scleritomes composed of multiple sclerites. In the Late Cambrian, however, a mollusk fossil interpreted as having a multiplated skeleton appears in the fossil record. *Matthevia* has been reconstructed as the earliest chiton. Chitons are slow moving mollusks with eight main articulated calcium carbonate valves. Modern chitons feed by rasping monerans off of rocks in the intertidal regions of seashores. *Matthevia* is thought to have lived in a similar fashion, and the reconstruction by Runnegar et al. (1979) shows *Matthevia* as a seven-plated chiton grazing on stromatolites (figure 4.26).

In addition to their eight articulated valves (figure 4.27), some modern chitons have numerous millimeter-sized calcium carbonate sclerites embedded in the muscular girdle surrounding the valves. These sclerites are arranged in a fashion that may be analogous to the arrangement of sclerites in the earliest Cambrian scleritomes discussed above.

Of the earliest Cambrian shelly fossils, many groups are truly problematic in the sense that not only do we have no idea what kind of animal made them, but also we have no clear conception of the function or functions of the skeletal remains. One shelly fossil of this type is *Microdictyon* (figure 4.28), a porous phosphatic plate

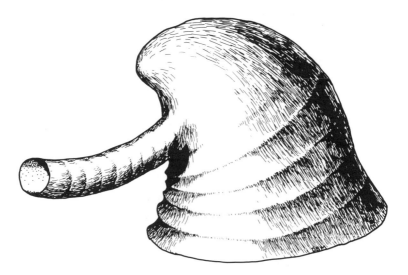

FIGURE 4.23. *Yochelsionella*, a Cambrian mollusk with a snorkel. Greatest width of specimen 3 mm.

with distinctive humps on the rims of the pores. As of yet, no one understands the nature or affinities of this mysterious netlike fossil (Bengtson et al. 1986). Other groups known from earliest Cambrian strata are equally mysterious; indeed, there is an anomalously high proportion of small shelly fossils that do not belong to later phyla.

"Living fossils" are creatures alive today that have undergone very little morphologic change for long stretches (sometimes 100 million years or more) of geologic time. Few living fossils remain from the earliest Paleozoic fauna. Monoplacophorans with small, cap-shaped shells linger as a remanent of the Cambrian fauna. Loz-

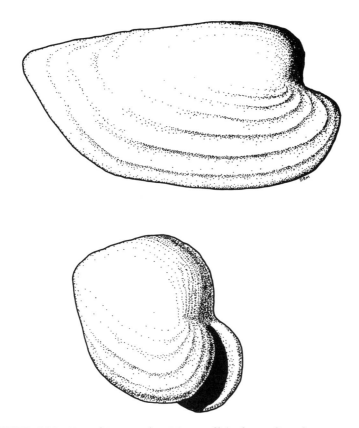

FIGURE 4.24. *Heraultipegma* (= *Watsonella*), the earliest known rostro-conch mollusk. Rostrochonchs superficially resemble bivalves, but their structure is quite different, as the two halves of a rostroconch shell are fused —they do not form separate valves. Shell approximately 2.5 mm long.

FIGURE 4.25. *Pojetaia runnegari,* a Lower Cambrian clam or bivalve. The earliest bivalves can be distinguished from rostroconchs by the articulations and ligament joining the valves, and by the muscle scars, where the muscles used in opening and closing the shell were attached. Width of valve 1.05 mm. (After Runnegar and Bentley 1983)

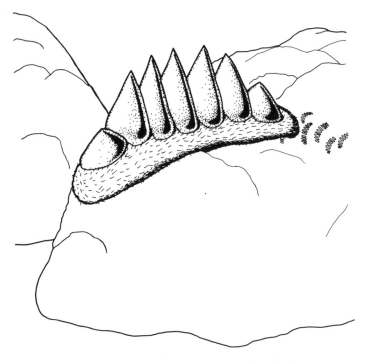

FIGURE 4.26. *Matthevia* (several centimeters in length) shown grazing on Upper Cambrian stromatolites. Compare this with the chiton in figure 4.27. (After Runnegar et al. 1979)

enge-shaped inarticulate brachiopods (such as *Lingula* and *Lingulella;* figure 4.19a) are a long-lived group that can be found today in sandy nearshore environments.

The only "living fossil" group from the Cambrian that went on to flourish in the modern world are the "seed shrimp" or ostracodes. Ostracodes are tiny (often about 1 mm in length) arthropods with a bivalved shell (figure 4.29) not unlike that of clams, except that it encloses seven pairs of tiny, jointed legs. Ostracodes are very abundant today both in modern marine and freshwater environments. The oldest examples are earliest Cambrian in age. Very early ostracodes are known from Estonia in the USSR, and belong to the phosphatic-shelled genus *Bradoria* (Mel'nikova 1987). The first calcareous shelled ostracodes do not appear until the end of the Early Cambrian (Koneva 1978). Superb preservation of ostracodes by phosphatization occurs in the Upper Cambrian black nodular limestones of Sweden. Müller (1981) has described ostracodes from these beds that show seven pairs of exquisitely preserved appendages.

The beginning of the Cambrian is marked by an impressive array of new animal types—this is the "Cambrian explosion" or "Cam-

FIGURE 4.27. The recent chiton *Chiton magnificus* from the Chilean coast. The head end of the animal is to the left. Note the numerous small sclerites along the margin of the eight major imbricate plates. Specimen 4.1 cm in length. (Specimen number 3586 of the teaching collection in Clapp Laboratory at Mount Holyoke College)

brian adaptive radiation event." It must be clear at this point that this event has nothing to do with bombs or nuclear radiation, but instead refers to the geologically rapid spread of new types of animals. The most important aspect of the earliest Cambrian faunas is simply this: representatives of nearly all major types or phyla of living animals appear during the first few million years of the Cambrian Period. The only well skeletonized and easily fossilized phylum of invertebrate animals that doesn't make its first appearance in the Cambrian is the phylum Bryozoa. Bryozoans are diminutive colonial animals related to brachiopods.

Many of the groups that were most important in the Cambrian

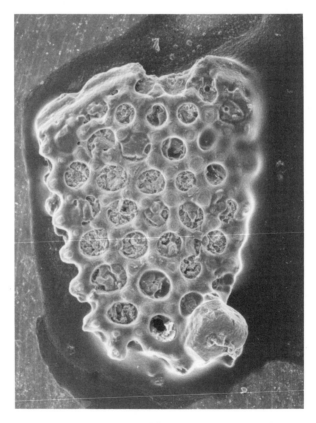

FIGURE 4.28. *Microdictyon*, a problematic Lower Cambrian phosphatic netlike shelly fossil, from the Puerto Blanco Formation Sonora, Mexico. Greatest width of specimen, 0.7 mm.

are unimportant or extinct today, for example, the trilobites, the inarticulate brachiopods, hyoliths, monoplacophorans, eocrinoids, the sclerite-bearers, and phosphatic tube-formers. True metazoans were undoubtedly present before the Cambrian, but they were all, with the exception of *Cloudina* (figure 4.2) and some *Cloudina*-like fossils (such as *Sinotubulites;* figure 4.30), soft-bodied. New types of soft-bodied animals appear in the Cambrian as well, but our understanding of these forms is restricted to rare finds of Cambrian soft-bodied fossils, which are even rarer than finds of the Ediacaran fauna.

The most famous Cambrian soft-bodied fossil locality is the Burgess Shale of British Columbia, Canada. A fine introduction to the Middle Cambrian Burgess fauna can be found in the book *The Burgess Shale* by Whittington (1985). The Burgess fossils reinforce the

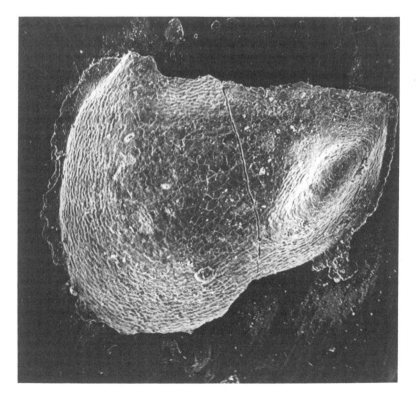

FIGURE 4.29. One valve of a Lower Cambrian phosphatic-shelled ostracode, from the Puerto Blanco Formation, Sonora, Mexico. Greatest width of specimen 2 mm.

FIGURE 4.30. *Sinotubulites,* a Lower Cambrian shelly fossil that is also known from the latest Precambrian, from La Ciénega Formation, Sonora, Mexico. Length of shell approximately 1.2 cm. (From M. McMenamin 1985)

pattern indicated by the occurrences of Early Cambrian shelly fossils, namely, that the Cambrian indeed witnessed an explosion of new animal types. The first priapulids (predatory worms with pentameral [five-sided] symmetry) and the first chordates are known from the Burgess Shale. The priapulid phylum exists today, and the chordate phylum includes animals with backbones such as fish and humans. We, and all the animals that share the surface of the earth and the seas, can trace our ancestry back to the Cambrian, but for most phyla the genealogy chart stops there. The trunk of our family tree disappears into a remote Precambrian past. And therein lies the central unresolved mystery of the origin of animals—why isn't there a fossil record in the Precambrian for the ancestors of most of the shelly animal phyla? We will return to this question in chapter 7.

The Golden Spike

IN GEOLOGICAL parlance, the word "Cambrian" can have two meanings. The first, and most commonly used, designates an interval of geologic time spanning from the second in which the first Cambrian sand grain was deposited until the instant that the first sediments of the Ordovician were laid down. No one was there with a stopwatch to mark these events, so these moments must remain theoretical instants of the geological past. Their importance is that they mark the beginning and ending of the sum total of Cambrian time. This time interval is called the Cambrian Period, and it is subdivided into the Early Cambrian, the Middle Cambrian, and the Late Cambrian Epochs. Periods and epochs are subdivisions of geologic time, and could have been measured with a stopwatch had anyone been there to time them. Radiometric dating of rocks can provide estimates of these times, but there are tens of millions of years of uncertainty with all radiometric dates of Cambrian age. It is not possible to know the beginning of the Cambrian Period with anything approaching stopwatch precision.

The second usage of "Cambrian" is as a designation of a particular body of rocks. These are all the rocks (both sedimentary and igneous) that were formed during the Cambrian Period. The term Cambrian System is used to refer to the sum total of rocks deposited during the Cambrian Period. With the help of a magic bulldozer, one could theoretically excavate all the patches of Cambrian rock worldwide and pile them into an enormous heap that would constitute the entire Cambrian System. One would need a legion of bulldozers to do this, because many surviving Cambrian rocks are buried quite deeply below the present-day surface of the earth.

The Cambrian System is divided into the Lower Cambrian, the Middle Cambrian, and the Upper Cambrian Series. The "Lower" and "Upper" used for series refer, naturally, to the physical positions of the strata from different parts of the Cambrian System. The Cambrian System is a "time-rock unit." In contrast to the Cambrian Period which is purely a time unit, a time-rock unit refers to all the rocks that formed within a certain period of time. Just as time units like "Cambrian Period" can be subdivided into epochs, the time unit "Cambrian System" can be subdivided into series. Series correspond exactly to the Early, Middle, and Late epochs introduced above. In this time-stratigraphic hierarchy, the "Early Cambrian Epoch" corresponds to the "Lower Cambrian Series." In other words, "Cambrian Period" and "Early Cambrian Epoch" are used in the same sense as "the year 1963," whereas the "Cambrian System" and "Lower Cambrian Series" are used in the same sense as "all the red wine bottled in 1963." The relationships between period, system, epoch and series are shown in figure 5.1.

Time Units		Time Rock Units
Era	~	Erathem
Period	~	System
Epoch	~	Series
Age	~	Stage

Rock Units

WESTERN TERMS	SOVIET TERMS
Group	Svit (Suite)
Formation	Gorizont (Horizon)
Member	
Bed	

FIGURE 5.1. Comparison of time, rock, and time-rock units. Both Western and Soviet rock terms are shown.

There is one problem regarding the recognition of the beginning of the Cambrian Period and its time-rock equivalent, the base of the Cambrian System. Neither the instant of time nor the point in the layered sedimentary sequence has yet been formally defined. A great deal of effort has been expended in recent years trying to locate the "best" point in rock to define the base of the Cambrian. Criteria for what is "best" include the ability to locate in other stratigraphic sections a point which is the *same age* as the point located in the formally defined standard section or stratotype section. When a stratotype point and section is finally agreed upon, geologists will be able to "drive the golden spike," and thus formally define the Precambrian-Cambrian boundary. The definition of a stratotype point is merely a formal, legalistic variant of the technique that resulted in the breakthrough in correlation made by European geologists in the 1830s (discussed in the first chapter). The golden spike is very useful, nevertheless, as the objective criterion against which a stratigraphic boundary can be correlated the world over.

The Cambrian System was a product of the pioneering geological research of the 1830s. Its initial definition involved two great early stratigraphers, Adam Sedgwick and Roderick Murchison. These British geologists began their studies of ancient fossil-bearing rocks in friendly collaboration, but their relationship deteriorated into a bitter feud over a terminological dispute. Secord (1986) has written an analysis of this tempestuous episode in Victorian science.

Using the new and powerful technique of correlating strata with fossils, Sedgwick and Murchison both began to study the stratigraphy of some of the most ancient sedimentary rocks known in Britain. At the time, these rocks were referred to as the "transition" rocks. "Transition" refers to the idea current before the 1830s that many of the most ancient sedimentary rocks known were transitional between the crystalline, igneous "primary" rocks (then thought to be the most ancient rocks on earth) and the "secondary" rocks that contained abundant fossils. Sedgwick began his work in North Wales; Murchison began his in the Welsh Borderlands. As this research progressed, Sedgwick named his rocks Cambrian; Murchison designated his, Silurian. The scientists agreed to a boundary between their two systems (with the Cambrian being the older system), and in 1835 they jointly published a short paper for the British Association for the Advancement of Science entitled "On the *Silurian* and *Cambrian Systems*, exhibiting the order in which the older Sedimentary Strata succeed each other in England and Wales" (Sedgwick and Murchison 1835). With continued study, it became clear that

Sedgwick's Upper Cambrian was equivalent to Murchison's Lower Silurian. Murchison was the more aggressive of the two, and, in an undiplomatic move, he tried to subsume Sedgwick's Cambrian into his Silurian. To Sedgwick's profound dismay, the newly created British Geological Survey adopted Murchison's classification of the disputed interval.

For many years, Sedgwick expressed his unhappiness with the situation to other prominent scientists. Much of Sedgwick's problem resulted from the fact that the Cambrian (in modern reckoning) rocks of North Wales are very poorly fossiliferous. This made it difficult for Sedgwick to demonstrate a distinctive fauna for the Cambrian. Stubblefield (1956) has shown that much of Sedgwick's original Cambrian System in Wales belongs to the Precambrian or to younger systems. As currently defined, the Cambrian in Wales consists primarily of fine-grained sedimentary rocks which have yielded only sparse upper Lower Cambrian faunas.

Some of Sedgwick's contemporaries favored retaining the term Cambrian, and others opted for abandoning it and, following Murchison's urgings, using Silurian to refer to all the oldest animal fossils. In the first edition of *Origin of Species,* Charles Darwin refers to the most ancient strata with animal fossils as Silurian (Darwin 1859). In a letter (dated May 30, 1857) to Sedgwick, the great anatomist Richard Owen (the scientist who gave dinosaurs their name) wrote:

> I have stuck to the "Cambrian" more instinctively, than through cold conviction, being bothered by the diversity of flat assertions as to fossil evidence from men whom I felt to be my masters in regard to that. In an old Diagram used a score of years ago nearly, at the Coll. of Surgeons, the base "Cambrian" supported, next, "Silurian." I used to be appealed to, to do away with the fundamental name, but could not make up my mind to it, and I hoisted the old flag again at my "Lectures," this year, in Jermyn Street.
>
> I began to think . . . that "[Cambrian] will come to be thought" all right, after all. But as "America" will never now be called "Columbia" . . . you will probably have your name attached to a very small part of your discoveries (S. Rachootin, personal communication, 1987).

With the passage of time, "Cambrian" was retained and Sedgwick was at least partly vindicated. "Primordial" fossils such as the trilobites described by the American geologist Ebenezer Emmons (1847;

see chapter 1, figure 1.2) were discovered at stratigraphic levels well below those at which anything had been found before, and were eventually assigned to the Cambrian. Interestingly, Emmons was quite correct in his assessment of the age of the New York trilobites, and he even went so far as to correlate the Taconic trilobites to the European system "known under the term *Cambrian* (p. 48)." In a passage that seems to predict the outcome of the Cambrian-Silurian dispute, Emmons challenged Murchison's claims that the Silurian contains the oldest known fossils:

> peculiar fossils [occur] on both sides of the Atlantic, which, so far as discoveries have yet been made, are confined to the slates of the Cambrian . . . system; and now the great object of the writer is to show that the above question [i.e., Murchison's proposal to call the oldest fossils Silurian] has not been settled right, or according to facts; or, in other words, . . . all the Cambrian rocks are not Lower Silurian (1847:48–49).

When a major unconformity (gap in the sedimentary record) was recognized between Murchison's Lower and Upper Silurian, many geologists switched to a three-part classification of this part of the geologic time scale, with the middle part often being referred to as "Lower Silurian." The Silurian-Cambrian dispute ended in 1879 when Charles Lapworth proposed the Ordovician Period and System for the contested strata. By the turn of the century, the terms Cambrian, Ordovician, and Silurian were in wide usage, and they remain today as the universally used lowermost three periods or systems of the geologic time scale. The basic, period-level geologic time scale has not changed since the term Ordovician was proposed.

More recent work has led to calls for the creation of a sub-Cambrian period and system. Abundant Ediacaran fossils occur in the western Soviet Union, and Soviet geologists have proposed the term "Vendian" as a period before, and a system below, the Cambrian (Sokolov 1952; Sokolov and Fedonkin 1984). A counterproposal was offered by Cloud and Glaessner, suggesting that this interval be called the Ediacarian Period and Ediacarian System (note the additional "i" in this spelling), with a type section in the Flinders Ranges of South Australia. Cloud and Glaessner discount the Vendian System because the stratotype section for the Vendian is known only from borehole cores, and is "inaccessible to direct observation" (1982:791). It is true that accessibility for collecting and study is one of the criteria for a suitable stratotype, but in our opinion this should

87

not affect the priority of the Vendian System. It makes good sense to find a Vendian stratotype that is more accessible than the original borehole sequences, but we don't think the term needs to be replaced with the unwieldy "Ediacarian."

Rowland (1983) has examined some terminological implications of the Vendian/Ediacarian dispute. For many decades, introductory geology students have been obliged to memorize the periods of the geologic time scale (figure 1.1), a duty that many of them found, and continue to find, onerous. One strategy for memorizing is a mnemonic aid that uses the first letter of every period to create a clever sentence. Some professors offer an award for the most clever and original mnemonic devised by student in each year's historical geology course. This contest almost always results in the creation of hilarious and innovative mnemonics. (Some entries cannot be read in public.) Many students fall back on old standard mnemonics, such as "Carl's Old Shirt Doesn't Match Pete's Pants" for Cambrian, Ordovician, Silurian, Devonian, Mississippian, Pennsylvanian, Permian. According to Rowland, with the adoption of a Precambrian period, the old mnemonics (some of which have been in use for over 100 years) must go.

> With Carl's Old Shirt headed for the trash, what's the next generation of geology students supposed to do to keep the periods straight? If Ediacarian prevails, try this as an aid: Every Class of Students Detests Memorizing Pointless Periods. Or, if Vendian is the choice: Very Cold Or Snowy Days May Prompt Pneumonia. (1983:82)

Regardless of the outcome of the latest terminology debate, there is as yet no formally defined Precambrian-Cambrian boundary. Both practical and philosophical approaches have been applied to the problem of defining the beginning or the base of the Cambrian. Some earth scientists have philosophically argued that all metazoan body fossils visible to the naked eye should be included in the Cambrian (Cloud and Nelson 1966). The unlikelihood of finding the earliest metazoans visible to the naked eye as fossils poses practical difficulties for pinpointing a correlatable boundary in sedimentary rocks. They were probably small, as well as soft-bodied. On the other hand, practical solutions to the Cambrian lower boundary problem that permit the presence of common and easily recognized fossils such as trilobites above and below the boundary (Bjorlykke 1982) seem overly pragmatic. With Bjorlykke's approach, fossils which many paleontologists would consider as being characteristically Cambrian would have to be assigned to the Precambrian. Bjorlykke makes an important point, however, because first appearances of fossil groups are not a suitable basis for a major boundary.

The concept of a Lower Cambrian Stage and Series was developed by continued application of the powerful (and pragmatic!) principles of biostratigraphy in the works of Barrande (1852–1911), Brøgger (1886), and Lapworth (1879, 1888). Walcott (1891) was first to divide the Cambrian into lower, middle, and upper series and his lead has been followed ever since. Walcott also (1889, 1890) succeeded in correlating the unique successions of Cambrian faunas with fossiliferous strata in North America, following the precedent of trans-Atlantic correlation set by Emmons (1847). Walcott's correct (1890) recognition of the Lower Cambrian trilobite fauna containing the trilobite *Olenellus* in North America was the first instance of successful intercontinental correlation of demonstrably Lower Cambrian sediments.

Following Walcott's (1890, 1900) success with intercontinental correlation, the *Olenellus* Fauna became widely accepted as a formal biostratigraphic zone, synonymous with the concept of the Lower Cambrian Series. But it soon became apparent that shelly fossils extended below the range of the *Olenellus* Fauna. G. F. Matthew was probably the first person to realize that there are shelly fossils stratigraphically below the abundant trilobites and other typically Cambrian fossils of the *Olenellus* Fauna (E. Landing 1987, personal communication). Matthew referred to this stratigraphically low interval (with few or no trilobites) of shelly-fossil-rich strata as the "Etcheminian." Matthew was not pleased when Walcott (1900), in a Murchisonian move, tried to include this low interval within the *Olenellus* Fauna (Matthew 1900). Matthew noted (1900:256) that "tube worms and brachiopods seem the most striking fossils of this lower" stratigraphic interval, and he was justifiably upset when Walcott (1900) tried to include this lower interval in the *Olenellus* Fauna without having demonstrated the presence of appropriate trilobites at this stratigraphic depth.

By the 1940s, in confirmation of Matthew's (1900) views, it was recognized that Precambrian-Cambrian boundary sections in many parts of the world possessed faunas of small shelly fossils underlying the lowermost trilobite-bearing beds. For example, Howell et al. (1944) called this interval the *Obolella* zone (*Obolella* is a small, oval-shaped inarticulate brachiopod). The global biostratigraphic importance of small shelly fossils was not appreciated until the late 1960s, when V. V. Missarzhevskii, A. Yu. Rozanov, and others began presenting the results of their studies on small shelly fossil faunas occurring before the first trilobites in the central and western USSR.

Unlike Matthew, who considered the early shelly faunas to be "distinct from the Cambrian" (1900:255), the Soviet workers in the 1960s assigned most of the early, pretrilobite shelly faunas to the Cambrian. This research culminated in the publication of the "brown bible" (Rozanov et al. 1969). In this book, the Tommotian Stage was defined, and the type section for the stage was along the banks of the Aldan River area in southeastern Siberia. The "brown bible" uses a combination of archaeocyathan and small shelly fossil range data to biostratigraphically subdivide the bottom of the Siberian Cambrian sequence into a number of biostratigraphic zones.

Faunas comparable to those of the Siberian Platform are now known from Scandinavia, many parts of the USSR, Poland, England, France, Newfoundland, Nova Scotia, India, Iran, New England, northwestern Canada, California and Nevada, northwestern Mexico, Mongolia, China, and Australia. Many of these faunas have been referred to as "Tommotian shelly faunas," although precise correlation with the Tommotian Stage of the Siberian platform is often questionable. It is in this sort of situation that the need for a "golden spike" is most apparent.

The International Union of Geological Sciences—International Geological Correlation Programme (IUGS—IGCP) Working Group on the Precambrian-Cambrian Boundary is commissioned with trying to locate a stratotype boundary for the base of the Cambrian. The three best candidate stratigraphic sections are sections: (a) along the Aldan River in Siberia; (b) in the Yunnan Province, Peoples' Republic of China; and (c) at Fortune Head in southeastern Newfoundland. None of these sections is completely satisfactory as a stratotype section. Both the Chinese and Soviet sections are rich with shelly fossils, but the sections are thin and are marred by significant gaps, manifest as unconformities. The Newfoundland section is thicker and the preserved sediments potentially represent much more time than is represented by the Chinese and Soviet sections (Signor et al. 1988). But the Precambrian-Cambrian boundary in Newfoundland is defined solely on trace fossil data (Narbonne et al. 1987), and the lowest shelly fossils occur more than 400 meters above the proposed boundary.

There are political aspects to the selection of a stratotype bound-

FIGURE 5.2. Canadian geologist Guy Narbonne points to the proposed Precambrian-Cambrian stratotype in a sea cliff at Fortune Head, southeastern Newfoundland.

ary. The boundary should be politically, as well as geographically, accessible. The Canadians, Soviets, and Chinese are vying with one another for the stratotype "prize." After all, who wouldn't want to have the "golden spike" in one's own geologic backyard? (Figure 5.2 shows a Canadian geologist eagerly promoting the Newfoundland boundary.) This stratotype boundary is the most important one in the entire geologic time scale, and once defined, all questions concerning the age and correlation of the Precambrian-Cambrian boundary must be referred to the sedimentary exposure into which the "golden spike" has been "driven." No doubt some enthusiastic stratigrapher will wish to erect a sizeable monument on the hallowed spot.

The Rifting of Rodinia

ANATOMICALLY MODERN humans first appeared on earth sometime during the "Ice Age" or Pleistocene Epoch of the Ceno-
zoic Period. There is no good reason to believe that this episode of
glaciation is over, and it is entirely plausible that the great North
American ice sheet could return in another ninety thousand years or
so and grind New York City into the Atlantic seafloor. There is an
interesting parallel between the origin of modern humans and the
emergence of animals—both events are associated with glaciation.
The first animals appeared during a long, drawn-out episode of glaciation at the end of the Precambrian.

Evidence for glaciation during the late Precambrian consists mainly
of unusual sedimentary deposits called tillites. Most sedimentary
rocks are composed of sediment particles that are more or less the
same size. For instance, sandstone consists primarily of sand grains,
or grains between one sixteenth of a millimeter and two millimeters
in diameter. If the sandstone consists almost solely of sand grains, it

is called a "clean" or "pure" sandstone. If the sand grains are all close to being the same size (a half millimeter in diameter, for instance), the sandstone is said to be very well sorted. Sorting of sediments occurs as a result of the winnowing action of wind and water. For example, light, tiny sediment particles can be carried off by winds and deposited together, leaving behind a lag of coarser grains too heavy to be moved by the wind. This natural segregation of sediment sizes by the action of wind and water is called sorting.

Tillites are the antithesis of a clean, well-sorted sandstone, because tillites include sediment sizes ranging from microscopic clay particles to meter-sized boulders mixed together in a helter-skelter fashion. The lack of sorting in tillites is due to deposition by glaciers. Glaciers carry a lot more than just ice as they move downhill; sediment of all sizes gets entrained in the ice mass. And since ice is a solid instead of a fluid or gas, it is unable to winnow or sort sediment the way wind and water can. When the edge of the glacier begins to melt (as when, for instance, it reaches the sea), the sediment of all sizes that was caught up in the glacier is unceremoniously dumped. This glacial debris, or till, is deposited in a pile or sheet of poorly sorted material. Assuming that the glacier is not enlarging or shrinking, large mounds of till can be deposited at the margin of the glacier. This can occur while the glacier remains at about the same size, since ice that melts at the glacier's foot is replenished by fresh snowfall at higher altitudes or latitudes. If subsequently lithified to sedimentary rock, this till becomes tillite, and can be preserved in the rock record as evidence for an ancient glaciation.

There is widespread evidence for glaciation in late Precambrian times. The rock record from one billion years ago to the Precambrian-Cambrian boundary is replete with rocks of known or suspected glacial origin, and it is certain that during this interval there were times when many parts of the world experienced exceptionally low temperatures. Harland (1983) cites evidence for at least four major episodes of late Precambrian glaciation. This 400 million year "reign of glaciers" is unique in earth history, and our Pleistocene glacial interlude is insignificant compared to the duration of the glacial periods in the late Precambrian. Never before or since have glaciers been so prevalent or frequent. The Varangian ice age (about 700 to 600 million years ago; the third late Precambrian glacial episode by Harland's 1983 rekoning) was the most severe climatic event of the Vendian, and Varangian tillites are widely distributed in

Svalbard, Greenland, Norway, Sweden, Scotland, Ireland, the western Soviet Union, and the Appalachians (Hambrey 1983). The Varangian (sometimes called Varangerian) glaciation is named for the Varanger peninsula in northeastern Norway, which has a more or less complete stratigraphic record from the Proterozoic to the Early Cambrian (Vidal 1984).

Ever since cartographers began making accurate world maps, children and other people with open minds have noticed that the east coast of South America and the west coast of Africa could fit together quite nicely. In fact, the fit seems to be so good that the two contients might be neighboring pieces of a jigsaw puzzle. With the acceptance of the theory of plate tectonics, it became clear that the jigsaw metaphor is indeed appropriate for describing the face of the earth (Monastersky 1987). Alfred Wegener, the now famous German meterologist and geophysicist who was scorned in his day for his support of the continental drift theory (Wegener 1967, translation of 1929 edition), surmised that today's continents were once connected to form a single supercontinent in the remote geological past. Wegener named this supercontinent Pangaea (from the Greek for "all land"). Had he not met an untimely death in 1930, the geological community might not have had to wait until the 1960s for widespread acceptance of the fact of Pangaea's existence. Virtually all geologists now accept that Pangaea, and its corresponding superocean Panthallassa, was formed about 200 million years ago by the coming together and collision of all or nearly all of the continents and continent fragments covering the face of the earth. The sea floor crust that existed between the pre-Pangaea continents was destroyed by subduction (underthrusting and melting) beneath the converging continents. After its formation, Pangaea was subsequently sundered by tectonic forces that are still not fully understood. Pangaean fragments were scattered over the earth's surface as new ocean basins (such as the Atlantic) formed in the fractures that split the supercontinent.

In the 1960s, the geological community was galvanized by geophysical evidence supporting continental drift, and as part of this "plate tectonic revolution," evidence became available indicating that the modern ocean basins may not have opened up only once (during the fragmentation of Pangaea), but two or more times during the course of geological time. In a famous article entitled "Did the Atlantic close and then reopen?", the Canadian geologist J. Tuzo Wilson suggested that the split between North/South America and

Europe/Africa was merely a rebirth of an ocean that was destroyed by the formation of Pangaea (Wilson 1966). Some scientists began to suspect that many ocean basins may have shared the same fate. If this were the case, it might imply the existence of a precursor super-continent that existed in remote Precambrian time, well before Pangaea. Evidence for a major, one-billion-year-old episode of continental collision and supercontinent formation began to accumulate in the early 1970s (Valentine and Moores 1972; Dewey and Burke 1973; Irving et al. 1974), and subsequently a large body of geological, paleontological, and paleomagnetic evidence has been marshalled in support of the existence of this Precambrian supercontinent (Lindsay et al. 1987). The supercontinent has been variously called "proto-Pangaea" (Sawkins 1976), "The Late Proterozoic Supercontinent" (Piper 1987), or simply the Precambrian supercontinent. The former existence of this supercontinent is now well established—in the late Proterozoic, there are unmistakable signs of its breakup (Conway Morris 1987a).

This supercontinent deserves its own name, so we here propose the name Rodinia (figure 6.1) for the Precambrian supercontinent and Mirovia for the corresponding superocean. These names are taken from Russian. A derivation from Russian seems fitting, be-cause of the important research done by Soviet earth scientists on the Late Precambrian and Early Cambrian (particularly their creation of the Vendian System). Mirovia is derived from the Russian word *mirovoi* meaning "world" or "global," and, indeed, this ocean was global in nature. Rodinia comes from the infinitive *rodit'* which means "to beget" or "to grow." Rodinia begat all subsequent continents, and the edges (continental shelves) of Rodinia were the cradle of the earliest animals.

A supercontinent need not include all the continents. Supercontinents have existed that comprised only a few or several continents. When Pangaea began to split apart, it initially broke into two main chunks, a northern supercontinent called Laurasia and a southern supercontinent called Gondwana. Laurasia consists of most of North America, Greenland, Baltica (primarily Europe west of the Ural Mountains), Siberia, Kazakhstania (southwestern USSR), and China, plus a few other minor continental blocks. Gondwana includes South America, Africa, Arabia, Iran, India, Madagascar, Antarctica, Australia, and New Zealand (figure 6.2). The rift between Laurasia and Gondwana formed a great east-west sea known as the Tethyan sea-way. Most modern oceans are elongate in a north-south direction, so

FIGURE 6.1. Rodinia, the Precambrian supercontinent, surrounded by the superocean Mirovia. The exact outline of Rodinia is unknown, but it approximately followed the continental edges exposed to Mirovia as shown in this figure. The South China platform may have fit in the gap between Baltica and India. (Reconstruction based on M. McMenamin 1982, Piper 1987, Donovan 1987, and Sears and Price 1978; created with TERRA MOBILIS)

the Tethyan seaway of 150 million years ago was unlike anything that exists today.

The makeup, as well as the breakup, of Rodinia is less well documented than that of Pangaea. There is good evidence suggesting that the southern continents were close together, in an arrangement that was very similar to that of the Paleozoic Gondwana. Evidence for this reconstruction is primarily from three sources: comparisons of Precambrian bedrock geology, paleomagnetics, and paleobiogeography.

Use of bedrock geology to reconstruct ancient continental positions relies on the idea that if two separated continents were once joined to form a single, larger continent, then there ought to be distinctive geological terranes (such as mineral belts, mountain chains, bodies of igneous rock of similar age, and other roughly linear to irregularly-shaped large-scale geologic features) that were once contiguous but are now separated. Matching of these features can provide clues to the positions of continents that were once together. Such evidence was used by Sears and Price (1978) to argue that Siberia was connected to the western (present-day coordinates) coast of North America during the Precambrian. The main problem with using bedrock geology features to match continental puzzle pieces together is that many of the potentially most useful linear geologic features on the continents (such as volcanic arcs or chains of volcanoes, and continental margin fold belts or parallel mountain chains formed by compression of strata) are *parallel* to the edge of the continent. Therefore, these features generally run parallel to rift fractures, and are less likely to continue and be recognizable on any continent that was once connected to the continent in question.

Paleomagnetic evidence is an important tool for the determina-

FIGURE 6.2. Gondwana, the supercontinent as it existed before approximately 250 million years ago. (Created with TERRA MOBILIS) Note: TERRA MOBILIS is a registered trademark of C. R. Denham and C. R. Scotese, *Earth in Motion Technologies*, 1987, 1988.

tion of ancient continent positions and for the reconstruction of supercontinents. Nearly all rock types, be they sedimentary or igneous, contain minerals that contain the elements iron or titanium. Many of these iron- and titanium-bearing minerals are magnetic. A familiar example is magnetite, an iron oxide that was used (under the name lodestone) to form the earliest compasses. In a compass, of course, the magnetized compass needle has a tendency to align itself with the earth's magnetic field. The magnetization of a crystal of a magnetic mineral (such as magnetite) is established immediately after the mineral crystallizes from a volcanic melt (lava) but before it cools below the Curie point temperature. Each magnetic mineral has its own specific Curie point. The Curie point of magnetite is 578 degrees centigrade, which—by way of comparison—is hotter than the melting temperature of pure lead (327.4 degrees C) but less than the melting temperature of pure aluminum (660 degrees C). As the mineral grain passes through the Curie point, the ambient magnetic field is "frozen" into the crystal and will remain unchanged until the crystal is destroyed by weathering or once again heated above the Curie point. This "locking in" of the magnetic signal in igneous rock crystals is the crucial event for paleomagnetism, for it indicates the direction of magnetic north at the time the crystal cooled (sometime in the distant geologic past for most igneous rocks). The ancient latitudinal position of the rock (and the continent of which it is a part) can be determined by measuring the direction of the crystal's magnetization. For ancient rocks, this direction can be quite different from the direction of present day magnetic north. A major assumption of paleomagnetic studies is that the position of the north magnetic pole has stayed more or less in the same place (with the exception of magnetic polarity reversals, in which the north and south magnetic poles switch places. Polarity reversals need not concern us here; see McElhinny 1979).

Sedimentary rocks can also be used for paleomagnetic determinations. As sediment particles settle to the bottom of the sea or lake or wherever they end up being deposited, any that are magnetic will have a tendency, like tiny compasses, to align themselves with the earth's magnetic field. If enough magnetic sediment grains are incorporated into the sediment, the sedimentary rock can retain a magnetic signal (called a primary remanent magnetization) that is as useful for paleomagnetic studies as are the magnetic remanences in igneous rocks. Relying mostly on paleomagnetic data, Piper (1987) has reconstructed Rodinia as a supercontinent consisting of all the major continents (figure 6.1). As noted above, Gondwana continents remain in essentially the same position as they were when Pangaea formed, which is not surprising since Gondwana remained intact between the breakup of Rodinia and the formation of Pangaea. Baltica is placed at one end of Rodinia, and North China is placed, with question, on the other end. There are some uncertainties regarding the exact placement of the continental pieces of Rodinia, most notably the position of China and the identity of the continent to the west (present-day coordinates) of North America (Siberia is shown in this position in figure 6.1, following the results of Sears and Price 1978). Despite uncertainty about the placement of some of the pieces of the jigsaw puzzle, the available paleomagnetic evidence favors the existence of a Precambrian supercontinent.

Paleomagnetic reconstruction is a form of geological analysis that is, unfortunately, fraught with uncertainties. The original magnetization is easily altered by weathering and metamorphism, and can confuse or obliterate the original magnetic signal. An inherent limitation of paleomagnetic reconstruction of ancient continental positions is that the magnetic remanence only gives information concerning the rocks' latitudinal position, and gives no clue as to the original longitudinal position of the rocks in question. For example, southern Mexico and central India, although nearly half a world apart, are both at about 20 degrees North latitude, and, therefore, lavas cooling in either country would have essentially the same primary magnetic remanence. One of the few ways to get information about the ancient longitudinal positions of continents is to use comparison of life forms on different continents. The study of ancient distributions of organisms is called paleobiogeography.

Debrenne and Kruse (1986) have determined that sixteen identical species of archaeocyathans are found in Lower Cambrian sediments of both Australia and Antarctica. Most archaeocyathan species have fairly limited geographic distributions, and Debrenne and Kruse (1986) state that the large number of archaeocyathans that are common to both Australia and Antarctica confirm the existence of a Gondwana supercontinent in the Early Paleozoic. These two continents are also likely to have been together in the Proterozoic. Because of their shared Cambrian biota, the Australian and Antarctican pieces of the of the Rodinia jigsaw puzzle are confidently in place with respect to one another.

Unfortunately, other Cambrian shelly fossils have so far proven less useful for determining the paleogeographic positions of conti-

99

nents, because they have very wide geographic distributions. Phosphatic tubular small shelly fossils such as *Anabarites* and *Hyolithellus* (figures 4.3 and 4.5, respectively) and trilobitoid trace fossils such as *Monomorphichnus* (figure 6.3) are extremely widespread on a global scale. Part of the problem with using these simple shelly fossils for biogeographic study is that there is very little information available for recognizing differences between species. Different biological species of *Hyolithellus* may have looked very different in life, but their fossil remains (consisting of a simple annulated tubular shell) may be indistinguishable. Some species of Early Cambrian shelly fossils do seem to be endemic, or restricted to certain geographic locales. *Lapworthella filigrana* (figure 4.8) is known from at least three widely separated localities in western North America, but, so far as we know, nowhere else (figure 6.4). There may have been geographic, climatic, or biological barriers that prevented this species of *Lapworthella* from spreading more widely.

Using Cambrian fossil distributions to infer the makeup of Rodinia gives a useful first approximation of Vendian continental posi-

FIGURE 6.3. *Monomorphichnus,* a trace fossil formed by the scratching action of trilobite legs. Lower Cambrian of Newfoundland; same specimen is figured in Narbonne et al. (1987). Scale bar is in centimeters.

tions of Australia and Antarctica, but it is better practice to use fossils of organisms that lived during the Vendian to make inferences about the positions of Vendian continents. I (M.A.S.M.) attempted to do this in 1981 by plotting the distributions of certain members of the Ediacaran fauna on a Cambrian paleogeographic base map (M. McMenamin 1982). My tabulation showed that although the frond-shaped members of the Vendian soft bodied fauna (such as *Pteridinium;* figure 2.2) had a global distribution, other distinctive members of the fauna (*Dickinsonia* [figure 2.5], *Tribrachidium* [figure 2.4] and several others, plus the African and Brazilian occurrences of *Cloudina* [figure 4.2]), seemed to be restricted to Baltica and Gondwana continents. I (M.A.S.M.) argued that the Ediacaran fauna first evolved and was most diverse in Gondwana, and that (contrary to conventional geologic wisdom) Baltica was as close, or closer, to Gondwana than to North America during the Vendian (M. McMenamin 1982).

This poses a problem for the reconstruction of Rodinia, because Baltica is generally thought to be very closely allied (in terms of Precambrian bedrock geology) to North America, a continent that has never produced the Gondwana/Baltica-type Ediacaran fossils even though Ediacaran fossils occur abundantly in several North American localities. One possible resolution to this problem was offered by the Piper (1987) reconstruction of Rodinia (figure 6.1), which brings Baltica and the eastern end of Gondwana relatively close together in southerly latitudes (Donovan 1987). The detailed positions of individual continental blocks within the Vendian supercon-

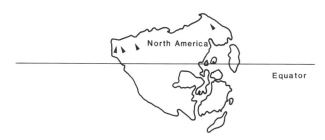

FIGURE 6.4. Distribution of the sclerite *Lapworthella filigrana* (illustrated in figure 4.8) on the Lower Cambrian North American continent. Triangles show localities where this species has been found. The straight line indicates the probable position of the Cambrian equator relative to ancient North America.

tinent may shift with continued research, but the Piper (1987) reconstruction of Rodinia seems to be a reasonable approximation.

Recent results suggest that rifting and continental breakup occurred throughout the world near the Vendian-Cambrian boundary, although episodes of rift-related volcanism occur well back into the Proterozoic. When a continent splits apart, it does so along a roughly linear fracture in the earth's crust called a rift basin. Rift valleys can be persistent features on the margins of a continent; the East African Rift is a currently active rift valley (one that has not yet opened up into an ocean). In initial stages of rifting, the rift basin fills with flows of a dark colored volcanic rock called basalt. Also filling the basin are rapidly deposited sediments washed in from the steep sides of the rift valley, as well as ash and stream-carried fragments of volcanic rock. If the rifting and tectonic tension that initiates the formation of a rift valley continues to fruition, a new sea will open up as the rifted halves of what was once one continent move away from each other, marine waters enter the deepening rift, and new sea floor basaltic bedrock becomes covered by normal marine sediments such as sandstones, shales, and limestones.

Lava flows interbedded with normal marine sediments are common in Vendian-Cambrian sedimentary sequences, and are known from Arabia, Mexico, the Ural Mountains, as well as other places (Bond et al. 1985; Zonenshain et al. 1985). In Sonora, Mexico, debris from ancient volcanic eruptions (found as a volcaniclastic conglomerate composed of basaltic cobbles and boulders) occurs in the stratigraphic section (figure 6.5) between the oldest trilobites known in Mexico (figure 4.14) and *Sinotubulites* (figure 4.30), a late Vendian to earliest Cambrian tubular shelly fossil. These volcanic beds may be related to an episode of continental rifting along the west coast of North America (Bond et al. 1985). Eruption of basalts sometimes indicates continental rifting, and in many Paleozoic sedimentary sequences in eastern and western North America, the basaltic volcanic intervals are in the Vendian or Cambrian parts of the section. This pattern has been interpreted to indicate that rifting around North America occurred around 600 million years ago (Bond et al. 1985). Continental rifting is always associated with eruption of basalts, and in many Paleozoic sedimentary sequences in eastern and western North America, the basaltic volcanic intervals are in the Vendian or Cambrian parts of the section. This pattern suggests that rifting around North America occurred approximately 600 million years ago (Bond et al. 1985). This interpretation is consistent with the observation that marine sediments deposited much before the Vendian in North America are generally confined to localized, restriced basins (Stewart 1976).

In the Vendian Tindir Group of Alaska and Yukon Territory (the extreme northwestern corner of the original North American continent), there is evidence for high-angle block faulting (Young 1982). Block faulting such as this is a common feature of continental breakup

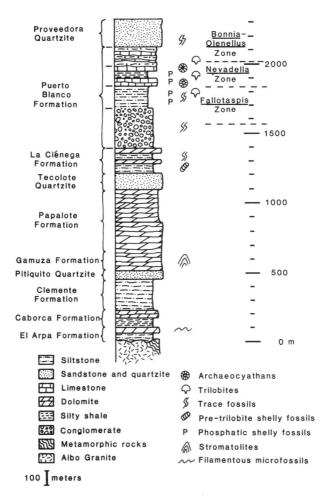

FIGURE 6.5. Precambrian-Cambrian stratigraphic section of the Caborca region, Sonora, Mexico. (After M. McMenamin 1984)

103

and rift valley formation. For example, the East African Rift Valley (including Olduvai Gorge, the site of numerous finds of early homonid fossils) is characterized by numerous high-angle block faults at the valley margins. Thus, there is ample evidence suggesting that at least the North American part of Rodinia was rifting away from the supercontinent cluster near the Vendian-Cambrian boundary.

The history of the supercontinent Rodinia can be summarized, in very general terms, as follows. Sometime during the latter part of the remote Precambrian past (about a billion years ago), this supercontinent was intact and was composed of most, or all, of the present day continental blocks or cratons. The exact shape of this supercontinent is unknown, but educated guesses as to its shape have been made using the best available paleomagnetic data. By Vendian time, approaching the end of the Precambrian, Rodinia seems to have been still largely intact, but there is evidence (rift basins, volcanic deposits on continental margins) that it was beginning to feel tensional forces that would eventually break it asunder. Limited paleobiogeographic evidence indicates that Baltica was near Gondwana, and in one reconstruction Rodinia has a more compact shape than the later supercontinent Pangaea. The Vendian saw four major phases of glaciation. We don't know if Rodinia had a large ice cap covering large areas of the supercontinent during the Vendian, but continental interior climate must have been extremely cold at times; supercontinents have very severe climates owing to the isolation of interior land from shoreline. Marine shorelines exert a moderating, maritime influence on climate, which is why Portland, Oregon has less severe winters than Minneapolis, even though Minneapolis is at a lower latitude.

Separate all the continents composing the supercontinent Rodinia (or Pangaea) and the total length of coastline is more than doubled, which would undoubtedly improve global climate. Therefore, if parts of Rodinia were at high latitudes or if global climate were severe, we would expect that many areas would experience glaciation. Another factor that could contribute to the Precambrian glacial record should be noted. During the early stages of rifting, the geothermal heat flow underneath the section of crust about to be rifted increases dramatically. Since hotter rocks are less dense, the crustal rocks, feeling the heat, rise up relative to surrounding, cooler areas of continental crust. As these areas are uplifted, they are carried into higher altitudes where the air is cooler and the chance that snow and ice will not all melt away during the summer months (a necessary condition for the formation of glaciers) is increased. Ironically, an increased heat flow from the earth's interior can cool the surface climate directly above it. This is in fact happening today—in the lofty Ruwenzori Mountains of Ethiopia (5119 m high), uplifted by the bouyant effects of East African rifting, glaciers are forming at a latitude that is practically on the equator! Glaciation attributable to this same rifting-related process of uplift may explain some of the Vendian glacial deposits (K. Bjorlykke, personal communication, 1981).

Near the Vendian-Cambrian boundary, parts of Rodinia (including North America, Siberia, Kazakstania, and probably others) began to rift away from the main supercontinental body, although Gondwana seems to have remained largely intact. Sea level, at an extreme low point by the end of the Vendian, began to rise at the Vendian-Cambrian boundary, partly in response to changes in the depth and volume of ocean basins worldwide. These ocean volume changes were, no doubt, at least partly a result of the creation of new oceans. It may seem strange that sea level could go up as a result of the creation of new ocean basins, but there is no contradiction here. For every new square kilometer of basaltic sea floor crust created between the edges of a rift basin, old (and cooler and less bouyant) basaltic crust has to be destroyed (assuming that the surface area of the earth has remained constant). The destruction of old seafloor occurs by subduction (underthrusting and melting) at deep ocean trenches; once melted, the old crust is returned to the surface in the form of volcanic eruptions. Ocean basins floored by old, cold crust can hold more water than those floored by newer, hotter basaltic crust because the cold crust rides much lower—owing to its greater density—and is able to "sink" deeper into the pliable part of the earth's interior that is found below the earth's crust. As Rodinia was cleft into separate continents, Mirovia gradually became divided into a network of smaller, shallower ocean basins as a result of the rifting.

Although the details of the breakup are still being worked out, it appears that the rifting of Rodinia was responsible for the volcanic beds in marine deposits, the opening of new ocean basins, a major increase in sea level, and possibly, high altitude glaciation at the uplifted margins of rift basins. A fifth effect was potentially even more important for the early evolution of animals. This was the introduction into marine waters of large quantities of biologically important chemicals and elements, partly as a result of rift-related volcanism. This phenomenon will be discussed in chapter 8.

VII

The Garden of Ediacara

Anatomy is destiny.
SIGMUND FREUD

E VERY BODY needs food. Less simply put, all living things require energy-rich molecules that can be broken down by organisms to provide energy. As mentioned in the introductory chapter, some autotrophic organisms are able to create these molecules by using energy sources and chemical building blocks taken from the non-living environment that surrounds them. The bodies of heterotrophs are dependent on autotrophs to fabricate these energy-rich molecules for them. Because photosynthesis is by far the most common process of autotrophy, nearly all of the food on earth is ultimately fashioned from the energy of sunlight.

Photosynthesis is generally considered to be a characteristic of plants in the traditional usage of the term "plant." Nonbiologists are sometimes surprised to learn that animals such as the blue dragon sea slug also are photosynthetic, as was discussed at the end of Chapter 1. One might argue that marine animals with zooxanthellae (symbiotic protists) are not truly photosynthetic because it is the protists that do the photosynthesis, not the animal. The protists just happen to be inside the animal. We would argue that this is not an important consideration, since photosynthesis in all eukaryotic (nucleated) cells is accomplished by chloroplasts, tiny organelles that are the cell's photosynthesis factories. Chloroplasts are now thought by many biologists to have arisen by a symbiosis event in which a small, photosynthetic moneran took up symbiotic residence within a larger microbe (Margulis 1981). The symbiotic relationship eventually became so well established that it became an obligatory relationship for both the host microbe and the smaller symbiont moneran. Reproductive provisions were made to pass the genetic material of the symbiont, as well as the host, on to succeeding generations. It would sound strange to describe an oak as a "multicellular alga invaded by photosynthetic moneran symbionts," but that is—in essence—what a tree is. Animals with photosynthetic protists in their bodies are able to create food internally, in the same way that an oak tree can, so we feel that these animals can be correctly called photosynthetic.

Trophic strategy can influence anatomy in important ways. The bodies of many multicellular animals have been modified to enhance their sunlight-capturing abilities. Guy Narbonne has suggested to us (1988, personal communication) that exclusively heterotrophic feeding may not be as basic an attribute of animals as one might expect. Many of the most primitive types of living metazoa contain photosymbiotic microbes or chloroplasts derived from microbes. Examples include the tiny green turbellarian worm *Convoluta,* the green hydra *Chlorohydra,* the sedentary jellyfish *Cassiopeia,* and the beautiful, green, sea anemone *Anthropleura* (Pearse et al. 1987). The blue dragon sea slug, with its symbiont-packed cerata, is yet another example of photosynthesis in an animal (Rudman 1987). The clam *Corculum* fabricates "windows" in its shell to admit more light for its internal protists (Seilacher 1972).

Flattened bodies with large, exposed surface area are also found in many animals with photosymbiotic protists. As noted earlier, Fischer (1965) first suggested that the flattened shapes of Ediacaran creatures would have helped photosymbiont tenants gain sufficient light. Seilacher (1984) championed this idea by claiming photosynthesis as a possible trophic strategy for these apparently gutless and mouthless Ediacaran creatures.

The most obvious reason for any organism, regardless of what kingdom it belongs to, to evolve a leaf-shaped body is to maximize its surface area. Leaf shape evolves in response to factors in addition to surface area requirement, but the surface area requirement, in all cases we are aware of, is the most important factor. Most of the Ediacaran frond fossils are leaf- or fern-frond-shaped—recall the derivation of the genus name *Pteridinium* (figure 2.2) from the Greek word for fern. Even *Dickinsonia* (figure 2.5) roughly resembles an elm leaf. Leaves of modern plants and Ediacaran animals probably evolved similar shapes for the same reason, namely, maximization of surface area.

It is important to emphasize that we are making an inference when we argue that the soft-bodied creatures of the Ediacaran fauna were maximizing their surface area for the purposes of autotrophic feeding. An inference in the sciences is a statement or assumption that cannot yet be unequivocally confirmed, but which seems to fit the facts available. As such, most scientific inferences have the same status as an educated guess in other forms of human discourse. An educated guess may seem a shaky foudation on which to base a major theory, but many breakthroughs in science have occurred when scientists were willing to gamble by looking at a scientific problem from a conjectural and unconventional perspective. With regard to the Ediacaran biota, it is certain that "normal" heterotrophic animals were present because of deposit-feeding burrows—clearly not all Vendian animals needed to maximize their surface areas. The unusual shapes of the soft-bodied fossils, however, require explanation, and the inference of photosymbiosis best accounts for the strange shapes of the Ediacaran body fossils.

Photosymbiosis is not the only possible departure from heterotrophic feeding, the usual method of food acquisition for modern animals. Seilacher (1984) notes that flat bodies are good for absorption of simple compounds such as hydrogen sulfide, needed for one type of chemosymbiosis. In chemosymbiosis as in photosymbiosis, microbes (in this case bacteria) are held within an animal's tissues as paying guests. The bacteria are able to use the energy stored in hydrogen sulphide molecules that diffuse into the host animal's tissues. The bacteria use the hydrogen sulfide to create food, using biochemical reactions that would be impossible for animals to do by themselves. The bacteria use some of the food for themselves, but great excesses are produced and passed on to the host animal's tissues.

The greatest zoological discovery of this century is that of the deep-sea vent faunas (Weisburd 1986 reviews a recent discovery).

These faunas include giant, gutless pogonophoran worms and clams. Such large animals are able to live in these lightless waters thanks to internal bacteria that metabolize the hydrogen sulfide percolating up from the volcanic activity of a mid-ocean rift as it creates new sea floor. The vent faunas appear to survive largely independent of sunlight-derived food, and the clams and pogonophoran worms have been termed "autotrophic animals" by Felbreck (1981).

There may be important similarities between the ecologies of these flattened Ediacaran creatures and the modern deep sea vent faunas. Most Vendian soft-bodied fossils are from sediments that were deposited in shallow well-lit marine environments, and photosymbiosis would have been easy for flat, soft bodied creatures living under these conditions. A few Vendian fauna localities, such as the Mistaken Point fauna of Newfoundland, were deposited in deeper water (Anderson and Conway Morris 1982). Assuming that they lived near where they were deposited, photosymbiosis would have been impossible for these creatures because sunlight does not penetrate to these great ocean depths. Must one then conclude that these organisms were all heterotrophic?

Almost none of the Mistaken Point soft-bodied fossils are also known from shallower water deposits. Of the soft-bodied, high surface area forms, only one has been found amidst shallower water Ediacaran faunas. One interpretation of these differences is that the Newfoundland soft-bodied creatures used their flat bodies, as Seilacher (1984) suggests, for the absorption of hydrogen sulfide or other nutritious gases. The presence of Ediacaran soft-bodied fossils in deep water sediments is therefore no proof that they were exclusively heterotrophic feeders.

The Mistaken Point fauna is deposited in a turbidite, a type of deep-sea sediment that would not necessarily form near sea-floor volcanic activity. If the organisms were indeed absorbing gases from the sea floor, this presupposes a constant source of gaseous nutrients. There is no evidence for hydrothermal activity in the Mistaken Point strata; where could such gases have come from? There are a number of possible sources for such gas. Sulfides can be produced in sediments by the microbial degradation of buried organic material, and can then slowly percolate to the sediment-water interface. Chemosymbiotic tube worms have been reported living over "breaches" or open rifts, in sediments off the Oregon coast (Anderson 1985). No hydrothermal activity or hot water springs are associated with the Oregon tube worms, and they seem to be feeding off of methane-enriched waters that are percolating up through the sediment rifts. There are tremendous reservoirs of methane and other natural gases held in sea floor sediment as gas hydrates (Kvenvolden and M. McMenamin 1980); these gases may leak to the sea floor over time, particularly after times of glaciation when the temperature of sea floor sediments is changing. Perhaps some or all of the Newfoundland organisms were trapping hydrogen sulfide, methane, or other nutrients that were escaping from sediments below, just as their shallow water counterparts were capturing sunlight radiating from above.

A form of chemotrophy (feeding on chemicals) that does not involve symbiosis is simple absorption of nutrients dissolved in sea water. Although this might not seem a particularly efficient way of obtaining food, there are tremendous amounts of "unclaimed" organic material dissolved in sea water. Monerans allow these nutrients to diffuse into their cells, a fact well known to microbiologists. Less well known is the fact that larger organisms can feed in this way also. Benthic foraminifera up to 38 millimeters long from McMurdo Sound, Antarctica, take up dissolved organic matter largely as a function of the surface area of their branched bodies (Delaca et al. 1981). These protists live under semipermanent sea ice. Even though this environment is usually poor in dissolved organic matter compared to other environments, these forams are able to satisfy their food requirements by the direct uptake of dissolved nutrients. Members of the Ediacaran fauna, particularly those without access to sunlight, may have also fed by absorption of dissolved nutrients.

Although there is as of yet no unequivocal proof, it seems reasonable to infer from their shapes that members of the Ediacaran fauna used photosymbiosis, chemosymbiosis, and direct nutrient absorption to satisfy their food needs. Since these methods do not involve killing, eating, and digesting other living things, we will refer to them as "soft path" feeding strategies. Heterotrophic organisms use "hard path" feeding strategies because they need to use up the bodies of other organisms for energy. The higher in the food pyramid, the "harder" the feeding strategy, on up to the keystone predator (top carnivore) at the top of any particular ecosystem's trophic pyramid. It is important to note that the term "hard," as used here, does not necessarily imply that autotrophic organisms have any easier a time obtaining their food than do heterotrophic organisms. Green plants are not very efficient at converting sunlight to food; sunlight can be thought of as an elusive prey because it is not a concentrated energy

source (Ricklefs 1976). Low food concentrations are a major difficulty encountered by organisms employing soft path feeding strategies.

Deposit feeding is intermediate between hard and soft paths. Much
of the edible organic material in sediments accessible to deposit
feeding organisms is dead, in the same sense that dissolved nutrients
in seawater are nonliving, although both were once derived from
living sources. Ingestion of nonliving organics in sediment is the
"soft" aspect of deposit feeding. This organic material, however, is
usually coated with bacteria and other heterotrophic monerans that
invariably get ingested and digested by deposit feeders, constituting
the "hard" part of deposit feeding. As noted earlier, trace fossils
indicate that deposit feeding undoubtedly occurred during the Vendian. Filter feeding, or capturing food suspended in the water, also
has components of both hard and soft paths because suspension
feeders can take both living and nonliving food from the water.

Many marine animals and protists today utilize photosymbiotic,
chemosymbiotic, or direct nutrient absorption feeding strategies.
Why, then, are flattened, soft-bodied forms with these food-producing habits rare in modern seas? Flattened forms are rare today, the
main modern exception being the fleshy, multicellular marine algae
that are both soft-bodied and reliant solely on soft-path feeding.

Glaessner (1984) puzzles over the reasons why soft-bodied organisms up to one meter in greatest dimension were able to colonize
the Vendian sea floor unmolested, and he notes that there is no
evidence for large predators in the Vendian. There is not a single
exclusively hard path organism known from the Vendian, particularly if one questions the jellyfish affinities suggested for some Ediacaran fossils. What are the implications of a predator-free Vendian?

One way to approach this question is by asking another question.
Have increases in the numbers, types, and abilities of predators
caused long-term changes in marine ecology? This question has led
to a proposal by G. J. Vermeij (1987) that the intensity of predation
has continually escalated throughout the last 600 million years,
resulting in improvements in the ability of prey to avoid being eaten.
Vermeij's hypothesis of evolutionary escalation can be viewed as a
positive feedback cycle, with ecological conditions within a given
habitat becoming more rigorous with the passage of hundreds of
millions of years. This view of predator-prey escalation as a dominant theme or trend in the history of life has been criticized as being
limited and suggestive rather than decisively convincing (Kohn 1987),

in part because of the difficulty of obtaining accurate paleoecological information from the imperfect fossil record. Nevertheless, an analysis of the paleoecology of Ordovician to Pleistocene brittlestar (a type of echinoderm similar to starfish) communities shows that long-term changes in predation pressure are likely to have been responsible for the demise or restriction of certain marine communities, particularly those dominated by filter feeders living near the sea floor (Aronson and Sues 1987). Brittlestars are sea-floor filter feeders that are very susceptible to predation by swimming predators such as fish. Since dense fossilized stands of brittlestars are less common after the Paleozoic than before, Aronson and Sues (1987) conclude that the frequency of brittlestar-dominated communites has decreased as a result of increases in the numbers and efficiency of swimming predators.

The hypothesis that the Cambrian skeletonization of animals was related to the appearance of predators has a long history, dating back over eighty years (Evans 1910; Schuchert and Dunbar 1933; Hutchison 1961). Glaessner (1984) dismissed this idea as simplistic and "anthropomorphic." The rarity of fossilized Cambrian predators encouraged arguments against the importance of predation in the Cambrian (Valentine 1973), in spite of Hutchison's (1961) suggestion that early predators were likely to have been entirely soft-bodied. Hutchison's inference has been largely borne out, thanks to recent analysis of Middle Cambrian soft-bodied fossils in the Burgess Shale of British Columbia, Canada. In addition to typical Cambrian shelly fossils such as trilobites, the Burgess Shale has yielded soft-bodied or lightly sclerotized predators such as the priapulid worm *Ottoia* and *Anomalocaris* (Whittington 1985).

Anomalocaris is a huge (by Cambrian standards) animal whose maximum length approaches half a meter. Its name means "anomalous shrimp," and was first applied to vaguely shrimplike segmented objects first recognized in the Burgess Shale and later found in the Lower Cambrian Kinzers Formation of Pennsylvania. These objects later proved to be *Anomalocaris'* paired frontal appendages. These appendages are usually found detached from the main *Anomalocaris* body. Also usually found separate is the circular mouth of *Anomalocaris*, which was originally described in error as the jellyfish *Peytoia*. Several complete specimens from the Burgess Shale have allowed Whittington and Briggs (1985) to combine the various fossil fragments and reconstruct *Anomalocaris* to its previous glory. The reconstruction of *Anomalocaris* (figure 7.1) resembles no living animal. The predator had a body shaped like a flattened teardrop, both sides of which were flanked by swimming fins. At its broad head were a pair of jointed appendages for drawing prey into its circular mouth, which was shaped like a pineapple ring, lined with teeth. *Anomalocaris* has been blamed for injuries to the carapaces of Lower and Middle Cambrian trilobites. These trilobites have gouges taken out of them, and the margins of the wounds display a raised rim, indicating the the the trilobite survived the attack and lived to heal the injury (Rudkin 1979). An unusual trilobite specimen—described as the species *Olenellus pecularis* by Resser and Howell (1938)—from the Kinzers Shale, is actually a specimen of the familiar Lower Cambrian trilobite *Olenellus thompsoni* with a damaged left side of the cephalon. Since *Anomalocaris* appendages occur in the same beds, it seems reasonable to infer that this predator was responsible for damage to the peculiar *Olenellus*.

Although examples of damaged prey and possible antipredatory adaptations greatly outnumber actual fossils of predators in the Lower Cambrian, some of the oldest known phosphatic small shelly fossils probably belonged to predators. Spines belonging to the genus *Protohertzina* (figure 4.6) are found in the earliest Cambrian strata, and, as noted earlier, have similarities in microstructure to the grasping spines of modern chetognaths or arrow worms. Arrow worms are a living phylum of voracious micropredators. They are free swimming and can attack prey under a few millimeters in length.

In addition to *Anomalocaris* and *Protohertzina*, there are a few other possible Lower Cambrian predator fossils. The Lower Cam-

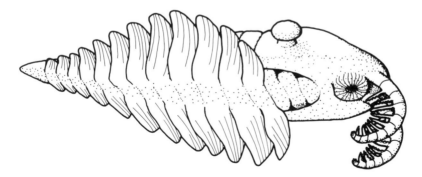

FIGURE 7.1. Reconstruction of the 45 cm-long Cambrian predator *Anomalocaris*.

113

brian trace fossil *Teichichnus* resembles the traces made today by mobile, carnivorous polychaete worms (Seilacher 1957). The trace-maker of *Dolopichnus*, a cylindrical burrow filled with fragments of trilobite carapaces, seems to have attacked Lower Cambrian trilobites (Alpert and Moore 1975). The *Dolopichnus* burrow may have housed a sea anemonelike predator capable of stunning trilobites and drawing them into its gut. An object from the Kinzers Formation described by Resser and Howell (1938) as a chela (claw) of possible crustacean origin has turned out not to be a chela, but may be a fragment of a predatory species such as *Anomalocaris* (D. E. G. Briggs, personal communication, 1986). Fossil chelae such as crab and lobster claws did not become abundant until well after the Cambrian, but D. H. Collins has recently discovered the fossil of a predatory arthropod, with five pairs of claws on its head, in the Burgess shale (Middle Cambrian). This new fossil is informally called "Santa Claws" (Collins 1985); this has been playfully echoed in the new formal taxonomic name for the species—*Sanctacaris uncata* (or "Santa Claws shrimp"; Briggs and Collins [1988]).

Some trilobites may have been predatory. Lower Cambrian specimens of the trilobite *Redlichia* from South Australia bear wounds that may have been inflicted by fellow trilobites (Conway Morris and Jenkins 1985). The Middle Cambrian trilobite *Olenoides* may have grasped small prey with the aid of spinose limbs (Whittington 1985).

Examples of possible antipredatory adaptations are common in Lower Cambrian faunas. If an animal can form a skeleton, forming spines as an extension of the shell is an effective way to deter predators. Some Lower Cambrian shelly fossils may have been devoted entirely to making a spiny protective coat, such as the *Lapworthella* (figures 4.8 and 4.9) scleritome discussed in chapter 4. The calcium carbonate sclerite *Chancelloria* belonged to an extremely spiny scleritome (Bengtson and Missarzhevskii 1981). The individual sclerites have six or more spines radiating from a central boss. Complete scleritomes (figure 4.10) of *Chancelloria* spicules are known from the Burgess Shale, and the *Chancelloria* animal must have been densely covered with spines.

Spines were in the defensive repertoire of other early shelly animals. The oldest articulate brachiopod, a specimen possibly belonging to the genus *Nisusia* (similar to the specimen shown in figure 4.20) from Lower Cambrian strata of the Siberian platform, has low tubercles on the shell exterior that have been interpreted as the bases of spines (Ushatinskaya 1986). Shell tubercles indicative of spines are well known from postCambrian brachiopods. The Lower Cambrian brachiopod *Acrothele spinulosa* from northwestern Africa has numerous slender, closely set spines radiating from the shell margin (Poulson 1960). Such spines could certainly serve a protective function.

One of the helicoplacoid echinoderms, *Helicoplacus curtisi*, bears spines on the upper surface of its spindle-shaped, plated exoskeleton. Although these spines would not be as effective as the spines on a modern sea urchin, the spines on this helicoplacoid would still have afforded additional protection to the upper surface of the animal (M. McMenamin 1986). If it was able to expand and contract its spindle-shaped skeleton, this helicoplacoid may have expanded when threatened, enlarging itself into a swollen object covered with short spines in a defensive strategy similar to that of a modern pufferfish or blowfish. In addition to spines, Cambrian echinoderms protected themselves by increasing the rigidity of their shell. Ridges and folds on eocrinoids are certainly protective because they make the echinoderm more rigid, and enable the inhabitant to live safely and "permanently inside the castle" formed by the calcite plates (Paul 1979:421). Later eocrinoids show greater degrees of rigidification than earlier ones. Helicoplacoids, the earliest known echinoderms (figure 4.18), may have gone extinct so soon because of an inability to rigidify their skeletal plates.

Many of the earliest known trilobites have elongate, pointed spines projecting from their carapaces. The "corners" or genal areas, of the head or cephalon of trilobites frequently developed into a pair of elongate genal spines. Long genal spines might have had uses as support on soft substrates (Clarkson 1979) or to aid molting (Robinson and Kaesler 1987), but they also would make the trilobite less susceptible to attack by predators (Clarkson 1979). Species of *Fallotaspis* (figure 7.2) and *Judomia* (figure 4.13) are representative early trilobites with very long genal spines. The trilobite *Callavia broeggeri* had a long projection—extending from the back of its cephalon—called the occipital spine (figure 7.3). The elongate genal spines and throracic spines of *Olenellus yorkense* (figure 7.4) made this trilobite less vulnerable to predators. Projecting genal spines and thoracic spines made it more difficult to flip a trilobite over and expose the softer underside. Although Lower Cambrian trilobites were unable to enroll (a common defense in Ordovician and later trilobites), even a slight flexure would have raised the spines into a

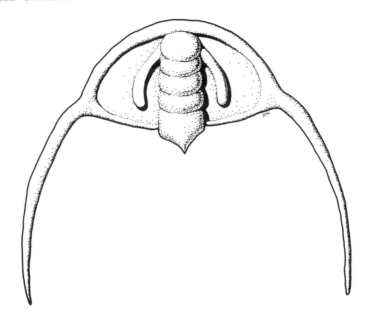

FIGURE 7.2. The cephalon of *Fallotaspis* (*above*), one of many Lower Cambrian trilobites with long genal spines. Width of specimen 5 cm. (After Harrington et al. 1959)

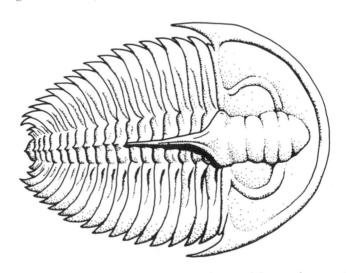

FIGURE 7.3. *Callavia* (*above*), a Lower Cambrian trilobite with a prominent occipital spine projecting from the posterior of the cephalon. Width of specimen 3.5 cm. (After Harrington et al. 1959)

116

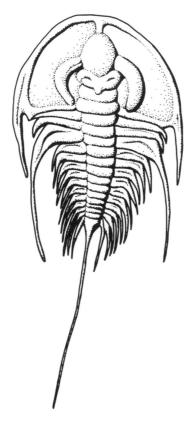

FIGURE 7.4. *Olenellus* (*above*), spiny Lower Cambrian trilobite. Length 5 cm. (After Harrington et al. 1959)

defensive posture. Some trilobites had to flex the cephalon downward in order to begin the molting process (McNamara and Rudkin 1984); this same reflex could have served a double duty by raising the genal spines into defensive position when the trilobite was threatened.

The Lower Cambrian trilobite *Laudonia* (figure 4.15) has conspicuous "extra" spines (called metagenal spines) projecting from the front part of the cephalic margin. These spines increase the effective maximum width of the trilobite in much the same way that erectly-held fin spines increase the diameter of many modern bony fish. Spines such as these are present in the larval stages of several related

117

trilobite genera, but only *Laudonia* retains elongate metagenal spines in adult stages. These spines may have dissuaded the contemporary predator *Anomalocaris* or other predators from frontal attack.

Spines, of course, are not the only strategy that prey can use to defend themselves against predators. Trilobites are the first organisms known to have had complex visual systems (Robinson and Kaestler 1987). A thorough search for eyes in the Ediacaran fossil *Spriggina* (a superficially trilobite-like form) has so far proved unsuccessful (Kirschvink et al. 1982), although if arthropods were present in the Ediacaran biota (Jenkins 1988), it won't be too surprising if Ediacaran body fossils eventually turn up sporting eye spots. Judging from the large, elongate eye regions (ocular lobes) of early trilobites such as *Judomia, Nevadia,* and *Fallotaspis* (figures 4.13, 4.14, and 7.2, respectively; M. McMenamin 1987b), the first trilobites had well-developed eyes that could have been used to avoid predators. Eyes can help animals that live on the sea floor to avoid being eaten, and well-developed eyes seem to have been rare or absent in the Ediacaran fauna.

Toxic secretions are also used by animals for protection. The inarticulate brachiopod genus *Mickwitzia* has dense punctae running through its shell valves (figures 4.19b, 7.5, and 7.6). These punctae have been interpreted (M. McMenamin 1986) as conduits for chemical deterrants used to discourage predators, because shells occurring in the same beds but belonging to different animals have been bored by parasites and predators (figures 7.7 and 7.8). Mickwitziid brachiopods, restricted to the Lower Cambrian, are some the the largest Lower Cambrian brachiopods known. A form from the Poleta

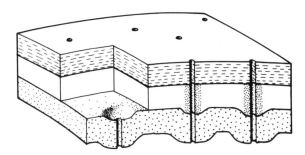

FIGURE 7.5. Shell structure of *Mickwitzia,* a Lower Cambrian inarticulate brachiopod (see also figure 4.19B). Note how punctae pass through all three shell layers. Width of valve 1 cm.

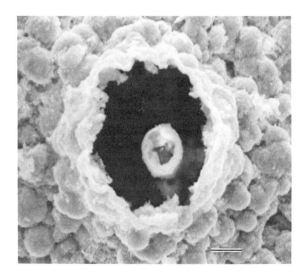

FIGURE 7.6. Enlarged view of a middle wall puncta in *Mickwitzia*, showing axial, hollow, phosphatic tube. The hollow tube may have carried chemicals to the exterior of the shell which were irritating or even toxic to boring organisms and other predators. From Lower Cambrian Puerto Blanco Formation, Mexico. Scale bar = 5 microns. (From M. McMenamin 1986; used with permission of the Society of Economic Paleontologists and Mineralogists)

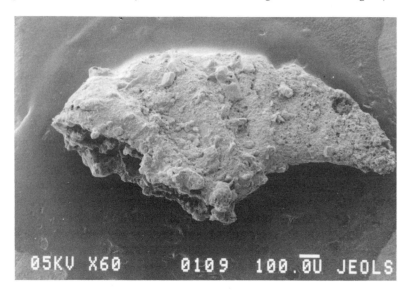

FIGURE 7.7. A specimen of the monoplacophoran *Bemella* with a possible bore hole near the apical end of the shell. Mickwitziid brachiopods (figure 7.6) found in the same beds have not been bored. From the Lower Cambrian Puerto Blanco Formation, Mexico. Scale bar = 100 microns.

Formation of California has a shell valve measuring 3.7 cm in diameter, a size that dwarfs most other Cambrian brachiopods (figure 7.9). The Lower Cambrian evidence presented above for antipredatory

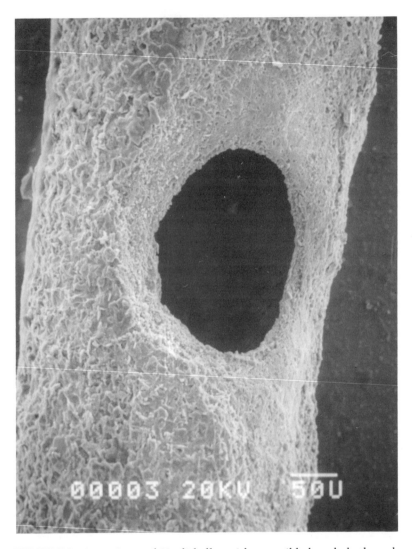

FIGURE 7.8. A specimen of *Hyolithellus* with a possible bore hole through the side of its shell. From the Lower Cambrian Puerto Blanco Formation, Mexico. Scale bar-50 microns.

defense and damaged prey stands in stark contrast to the lack of such evidence in fossil communities of the Vendian. The largest known organisms of the Vendian are the members of the Ediacaran fauna, and they are twice as large as the largest animals of the Cambrian.

FIGURE 7.9. The largest known Early Cambrian brachiopod; a large, partly crushed specimen of *Mickwitzia*. From the Poleta Formation of California. Greatest width 3.7 cm. (Photograph courtesy J. Wyatt Durham and Ellis L. Yochelson)

Despite the large sizes of the Vendian fossils, there is no known evidence for predation on any of the numerous Ediacaran fauna specimens. Indeed, it is a wonder that these soft bodied organisms existed on the sea floor unmolested (Glaessner 1984). The Ediacaran fauna may have gone extinct because it lacked defenses against Cambrian predators (Brasier 1979), although a few members of the fauna seem to have lingered on into the Cambrian.

The unusual fossil *Xenusion auerswaldae* (figures 2.1 and 7.10) was originally described in 1927 as an arthropod with multiple paired appendages. The type specimen of *Xenusion* was found in a slab of thinly bedded sandstone from surficial deposits of northern Germany. Jaeger and Martinsson (1966) traced the distinctive pinstriped sandstone of the *Xenusion* cobble back to the Kalmarsund Sandstone of southern Sweden, and presented a convincing case that the fossil was originally part of this Swedish formation and was subsequently

FIGURE 7.10. Side view of *Xenusion auerswaldae*, showing spines on medial humps. These spines are the only possible defense against predators that have been seen in an "Ediacaran-type" soft-bodied fossil. Spine length is conjectural because the spines are broken in all known specimens of this species. Length of frond approximately 7.5 cm. (From M. McMenamin 1986; used with permission of the Society of Economic Paleontologists and Mineralogists)

carried south across the sea to Germany by the action of Pleistocene glaciers. Jaeger and Martinsson (1966) gave *Xenusion* an earliest Cambrian age because of the occurrence of abundant vertical *Skolithos* burrows in the Kalmarsund Sandstone, and also agreed that this fossil was an organism with walking or swimming appendages. They made a latex cast of the fossil, and "flexed" the cast to give it a more "natural" (read "arthropod-like") appearance (Jaeger and Martinsson 1966; their figure 1, p. 437), perhaps because they were bothered by the fossil's "missing head" and its unarthropodlike concavity.

Halstead Tarlo (1967) reinterpreted *Xenusion* as allied to Ediacaran frond-shaped fossils such as *Rangea* and *Charniodiscus* (Glaessner 1979), in which case *Xenusion* would never have needed a head. Halstead Tarlo's (1967) interpretation seems most likely, particularly considering that other Ediacaran frond fossils have been confused with the thoracic regions of supposedly incomplete arthropod fossils. *Pteridinium*-like frond fossils from Stanly County, North Carolina (Gibson et al. 1984) were originally described as trilobite fossils lacking cephalons (St. Jean 1972). The individual "arms" in *Xenusion* are divided into eight or more segments, and all of these segments have faint parallel markings suggesting that they, too, may be divided into segments. Several orders of segment subdivision are seen in Ediacaran fossils such as *Rangea* (Jenkins 1985), and this feature accords with Seilacher's (1984) analogy between members of the Ediacaran fauna and air mattresses. *Xenusion* shares this "quilted" character, further supporting the hypothesis that *Xenusion* is a relict member of the Ediacaran fauna (M. McMenamin 1986). Note also the similarity of *Xenusion* to the Ediacaran 'pennatuloid organism' illustrated by Conway Morris (1989; his figure 2.4E).

Paired columns of humps run down the midline of *Xenusion* (figure 2.1), each of which bears an outwardly directed, prominent spine (figure 7.10). Because the only known specimens of *Xenusion* are incomplete, the original length of these spines is not known. A defensive function for *Xenusion*'s spines is plausible. They may, in fact, have been mineralized, although this cannot be proven with the fossil material currently available. Jenkins (1985) interprets the striate markings along the axis of one Ediacaran frond fossil as representing the impressions of stalk-supporting spicules. The spines of *Xenusion* may be stalk-supporting spines that grew at ninety degrees to the stalk axis, thus providing defense rather than support (figure 7.10). This type of defense would be compatible with an inferred soft path feeding strategy (such as photosymbiosis) in ways that an opaque, impermeable mineralized shell might not. On the other hand, such spines would not have been a particularly good defense. A new specimen of *Xenusion* has been found (again, from a glacially transported rock or glacial erratic; Krumbiegel et al. 1980; Schallreuter 1985) in which the secondary branches ("appendages") have been torn off and the central axis is contorted. This specimen was badly battered before being fossilized. Crashing waves might have damaged the new specimen in this way, but the injuries could also be the work of a marauding Cambrian predator.

Xenusion may be the last known member of an Ediacaran dynasty, a distinct period of earth history when flat or high surface area organisms (regardless of taxonomic affinites) had the sea floor to themselves, and survived using soft path feeding methods. Smaller predators were probably present at that time. Microfossils, known as heterocysts, reported from Vendian rocks, are thought to be the remains of heterotrophic protists (Bloeser 1985), and some Ediacaran animals were grazers and possibly filter feeders, but there is no evidence of predators capable of attacking the Ediacaran creatures. Why might this be the case? How could the trophic strategy of eating larger organisms remain undiscovered for millions of years? Unless there were large Vendian predators who have so far eluded paleontologists by avoiding fossilization or discovery, the time of large soft-bodied Vendian creatures can be called a largely predator-free "Garden of Ediacara." The uniqueness of this "garden" may go beyond a simple absence of large predators, however.

The Vendian-Cambrian transition records a profound turnover in the paleoecology of the marine biosphere, and this change is not solely expressed by the appearance of metazoa with biomineralized skeletons. Calcium carbonate-secreting algae appeared at the end of the Vendian (Riding and Voronova 1982), and later radiated to become important components of the earliest Paleozoic reefs.

A final example can illustrate the marine changes that took place during the Cambrian. Trace fossils belonging to soft-bodied organisms of enormous size (up to a half meter in length) are known from Upper Cambrian sediments deposited in very near shore, intertidal environments. One of these fossils, *Climactichnites* ("climax of trace fossils"), is nicknamed "Honda tracks" by paleontologists because it resembles the tracks made in sand by motorcycle tires (figure 7.11). *Climactichnites* is known from the Potsdam Sandstone of New York. The maker of *Climactichnites* was likely a mollusk with a very powerful crawling foot muscle. Also from the Late Cambrian are gigantic radular bites (tooth markings made by grazing mollusks; chitons [figure 4.27] can make these kinds of markings) from a tidal flat sandstone in Saudi Arabia. The creature that made these tracks was at least 10 centimeters in width (Seilacher 1977). Seilacher (1977:375) argues that these giant tracemakers were pioneers that left the water to graze monerans in an environment that was otherwise still uninhabited by higher organisms and "not yet endangered by terrestrial predators." In this last intertidal refuge of both large soft-bodied organisms and Garden of Ediacara monerans, metazoans seem to have been munching the monerans all the way back to the high tide mark.

If the decline of the Ediacaran fauna really was in part a result of the rise of Cambrian predators, the Vendian-Cambrian transition can be seen as a "fall" from the Garden of Ediacara. Quilted organisms such as the Vendian soft-bodied creatures are virtually unknown from later strata, suggesting that, in this case, anatomy was indeed destiny. This would be especially so if their soft path feeding requirements precluded the development of robust armor.

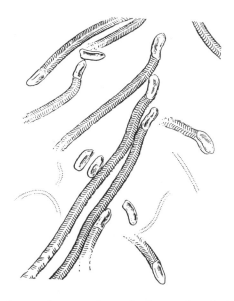

FIGURE 7.11. *Climactichnites wilsoni*, the largest described Cambrian trace fossil, from the Upper Cambrian of New York. Width of each trace fossil 10 cm; the tracemaker was a very large animal (perhaps up to one half meter in length) by Cambrian standards. (From Walcott 1912)

The fall from the Garden of Ediacara was a profound reorganization of our global ecosystem. What could have caused this transition from flattened, soft path feeders to a heterotroph-dominated, hard path biota? We will examine this question further in the next chapter.

VIII

Ecological Feedback and Intelligence

One of the most revolutionary biological innovations has been the development of the central nervous system.

A. G. FISCHER (1984: 153)

FEEDBACK

SCREEECH! GO the speakers and people in the auditorium slap their hands to the sides of their heads as the audio system spews out earsplitting noise. In this example of runaway positive feedback, the microphones were either placed too close to the speakers or the speaker volume was turned up too high. In either case, sound from the speaker enters the microphone, and the microphone causes the speakers to emit more sound, which causes more sound to enter the microphone, and so forth until someone turns the sound system off.

A particularly troubling example of positive feedback involves climate change caused by the use of chlorofluorocarbons (CFCs). CFCs are used as refrigerants, as aerosol propellants, and as solvents in the manufacture of high technology items such as computer circuits. Although safe for people, CFCs have proven to be very toxic for the atmosphere. As an unavoidable byproduct of their use, CFCs

127

escape in large quantities into the atmosphere. Ozone, or triatomic oxygen, protects organisms at the surface of the earth by absorbing harmful ultraviolet radiation. Some of this radiation has a frequency which will damage DNA molecules. Such damage can lead to fatal mutations, skin cancer, and other problems.

In addition to destroying the protective layer of atmospheric ozone, CFCs act as greenhouse gases. A greenhouse gas is any volatile chemical compound (carbon dioxide is another example) that traps radiant solar energy after it enters the earth's atmosphere. When sunlight, which passes easily through a greenhouse gas, strikes a dark surface, it is absorbed and reemitted as infrared radiation. With greenhouse gases in the atmosphere, this infrared radiation cannot pass as easily out through the earth's atmosphere as it could if the greenhouse gases were not present. This heat is therefore trapped near the earth's surface by an atmospheric blanket of greenhouse gases. The effectiveness of this blanket depends on the quantity and types of greenhouse gases in the atmosphere. All other things being equal, the more greenhouse gases in the atmosphere, the warmer the earth's climate will be.

Herein lies the positive feedback associated with CFCs. The more CFCs in the atmosphere, the warmer the climate will become. The warmer the climate, the more CFCs will be needed to run air conditioning units, refrigerators, and freezers. Also, as the protective ozone layer is depleted by CFCs, more people will be compelled to use sunblocks (sunscreen) to protect their skin from the damaging effects of solar ultraviolet radiation that passes in increasing amounts through the weakened layer of protective ozone. Some European countries market CFC-powered aerosol sprays of sunblock, the use of which depletes the stratospheric ozone layer further and increases the need for the product. Recently enacted CFC emmision control agreements are intended to limit such positive feedbacks (MacCracken 1987).

Feedback between organisms and their environment may be an important aspect of the global changes that occurred at the Vendian-Cambrian boundary. One of the effects of this paleoecological feedback is a series of extinctions that occurred during the Vendian.

VENDIAN EXTINCTIONS

THE EXTINCTION events of the Vendian tend to be overshadowed by the fact that the Vendian-Cambrian transition experienced the greatest radiation of multicellular life known. A mass extinction occurs when large numbers of organisms in different habitats die out at the same time, as occurred near the end of the Cretaceous, with the extinction of dinosaurs and all other large terrestrial animals, as well as many types of marine organisms. The earliest known, reasonably well documented, mass extinction is of Vendian age. There are indications of a cluster of mass extinctions in the middle to late Vendian, although the abruptness and simultaneity of these extinctions are somewhat obscured by the rarity of fossils and by the difficulty of obtaining precise dates for Vendian sediments. Also, some Vendian extinctions are the continuation of declines that began before the beginning of the Vendian, such as the loss of many different types of stromatolites.

Stromatolites (figures 1.3 and 1.4) reached a peak in diversity (nearly 100 recognized taxa; Awramik 1982) about 850 million years ago in the late Riphean. The Riphean is an interval of Proterozoic time used by Soviet geologists to designate an interval of geologic time before the Vendian; see figure 1.1. Following this acme, stromatolites underwent a precipitous decline, starting in the late Riphean and continuing through the Vendian. Stromatolite diversity bottomed out at fewer than 30 taxa by the beginning of the Cambrian. Although this decline does not necessarily represent the extinction of any of the individual moneran species that participated in the construction of the stromatolites, it does indicate that the conditions for many successful types of Proterozoic benthic microbial communities became much less favorable. For example, well-formed specimens of the distinctive stromatolite *Conophyton* (figure 1.3) are unknown after the Vendian. The advent of burrowing and grazing metazoans, and disturbance of microbial mats as a result of their activities, has been hypothesized as the factor responsible for the decline of stromatolites (Awramik 1982).

This decline as a result of overgrazing is an event of immense and global significance. Stromatolites were the communities that characterized the shallow water marine basins of the earth for three billion years, and were reduced in diversity to small numbers in a relatively short period of time (Sokolov and Fedonkin 1986).

Microbial species seem to be largely unaffected by the defeat of the stromatolites. It may be difficult, however, to recognize turnover of species in floras consisting primarily of simple round and tubular monerans. Similar looking microfossils may have had radically different genetic programming or biochemical machinery—it is impossible to tell this from the fossils alone, however. This problem is further compounded by the fact that fossilized sea floor microbiota are rare after the beginning of the Cambrian. We do know that certain distinctive microbial forms survived the Vendian without major change. The distinctive helically coiled (like a corkscrew) *Obruchevella* (Cloud et al. 1979), plus the earliest calcareous algae (Riding and Voronova 1982) crossed the Vendian-Cambrian boundary with impunity. *Obruchevella* has been recently discovered in silts of the Middle Cambrian Burgess Shale (Mankiewicz 1988).

Another Vendian sea-floor (or possibly planktonic) group that lives on into the Cambrian are the vendotaenid algae. Vendotaenids are ribbon-shaped, possibly moneran fossils, often microscopic but up to 1 mm in width and 10 cm in length. The first fossil evidence for primitive fungi (actinomycetes) are Vendian actinomycetes discovered in the process of decaying a fragment of a vendotaenid (Sokolov and Fedonkin 1986).

Acritarchs are another important group of Vendian fossils. Acritarchs are thought to primarily represent fossils of protists. They are organic-walled vesicles, frequently round and usually less than a tenth of a millimeter in diameter. They are studied by removing them from sediment by acid dissolution of the rock matrix. Acritarchs can be recovered from calcium carbonate rock (limestone) by dissolution with acetic acid, formic acid, or hydrochloric acid. More usually, however, the sediments containing acritarchs are silica-rich rocks such as a shale, and the more dangerous hydrofluoric acid must be used to dissolve the siliceous minerals (this technique is called maceration). Once removed, acritarchs can be studied by using transmitted light and scanning electron microscopes (Vidal 1984).

W. R. Evitt coined the term "acritarch" as a catchall category for small, organic, walled fossils of uncertain biological relationships. As such, the acritarch group can be though of as a "garbage can" term for microfossils that cannot be directly compared to known organisms. Often brown in color and collapsed or torn, acritarchs are not beautiful fossils; acritarchs recovered from maceration residues *look* just as though they came from a garbage can. Nevertheless, these fossils are crucial for biostratigraphy. Most acritarchs resemble the resting cysts made by unicellular, planktonic protists called dinoflagellates. Dinoflagellate blooms are responsible for the so called "red tides" that can have disastrous effects on fish living nearby. When a living dinoflagellate experiences unfavorable conditions for growth, it encloses itself in a resistant vesicle of a tough organic compound, called sporopollenin, and sinks to the bottom to await better conditions (Evitt 1986). By comparison to modern dinoflagellates, most acritarchs are thought to represent the resting stages of photosynthetic, free-floating protists. Acritarchs may be smoothly spherical, spiny (some look like miniature World War I-vintage floating mines), or octahedral (Vidal 1984).

The diversity of these planktonic microfossils underwent a severe decline during the middle to late Vendian, which a number of paleontologists accept as indicative of major extinctions in the eukaryotic phytoplankton (Vidal and Knoll 1983). Diagnostic acritarch species, such as the tongue-twisting *Trachyhystrichosphaera*, were gone by the end of the early Vendian. These late Riphean-early Vendian acritarchs (Yankauskas 1978) were replaced by a very low diversity planktonic flora typified by vendotaenids and the distinctive acritarch *Bavlinella* (figure 8.1). *Bavlinella* resembles living colonies of spherical cyanobacteria, and is probably some type of planktonic moneran. Planktonic monerans are outnumbered in today's seas by planktonic eukaryotes such as dinoflagellates.

The sediments containing this depauperate middle to upper Vendian acritarch flora have curiously large amounts of an organic material called sapropel. The sapropel is derived from burial of huge amounts of organic material. After this low diversity interlude, acri-

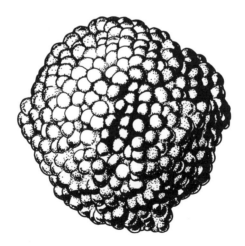

FIGURE 8.1. *Bavlinella faveolata*, an acritarch composed of multiple spheres, may have been a colonial prokaryote. This species is common between the middle Vendian and the latest Vendian. Width of specimen 25 microns. (After Vidal 1976)

tarch diversity did not recover to its early Vendian levels until well into the Lower Cambrian, when spiny forms such as *Skiagia* (figure 8.2) became abundant.

Pre-Vendian animals may have existed, but we know virtually nothing about them. The Riphean-Vendian boundary is about 700 million years old, a moment at—or just before—the start of the Varangian glaciation. Animal fossils from before this glaciation are very rare. A twisting string of what have been interpreted as animal fecal pellets is known from the upper Riphean of the Ural Mountains (Sabrodin 1972). As noted in chapter 3, the billion year old *"Brooksella" canyonensis* from the Grand Canyon has recently been reinterpreted as a complex trace fossil (Kauffman and Fursich 1983), although this interpretation is hotly contested by those who believe that it is an inorganic sedimentary structure (Cloud 1968). The oldest burrow convincing to us is a tiny, approximately 700 million year old, backfilled burrow from South China (Awramik et al. 1985). Also from China are 740 to 840 million year old annulated tubular structures that may be the oldest record of metazoa (Sun 1986), although the dating of these fossils cannot be accepted without reservations (Cloud 1986). Although the pre-Vendian animal fossil record is very scrappy, it seems plausible that grazing animals were present as far back as a billion years ago, to accord with the decline

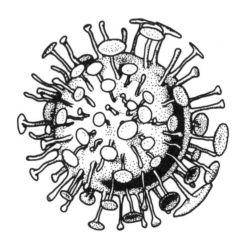

FIGURE 8.2. *Skiagia ciliosa*, a spiny acritarch known from the Lower Cambrian when acritarch diversity began to recover after the Vendian mass extinction of the marine phytoplankton. Width of specimen 24 microns. (After Knoll and Swett 1986)

in stromatolite diversity that began about 850 to 900 million years ago (Awramik 1982). It is tempting to speculate that the evolutionary development of these pioneer animals was delayed or halted by the severe Varangian glaciation, although there is not enough evidence at this time to make any claims for a late Riphean mass extinction event. Nonetheless, Sokolov and Fedonkin (1986) infer that the Varangian glaciation interval saw mass extinction of some groups of invertebrates of which we know nothing.

Survivors of this "great cold" diversified into the distinctive soft-bodied Ediacaran fauna. On the Russian Platform (where the Vendian was first recognized), the Vendian is divided into three horizons, the lower Redkino, the upper Kotlin, and the uppermost Rovno. Sokolov and Fedonkin (1986) see a rapid expansion of the fauna in the Redkino after the glaciation, followed in the Kotlin by an episode of extinction. Only rare problematic forms of metazoa and small trace fossils are known from the Kotlin Horizon of the Russian Platform. In the Rovno (uppermost Vendian), there is an abrupt increase in the dimensions of animals, as is indicated by the sizes of the trace fossils they left behind. Trace fossils became larger, more complicated, and deeper, indicating a greatly increased level of colonization of the sea floor by animals (Fedonkin 1978). Animals with resistant tubular skeletons appear in the Rovno. Sabelliditid tubes are common in the Rovno. These thin organic-walled (not mineralized) tubes presumably housed a worm-like, filter-feeding organism; the Vendian genus name *Sabellidites* recalls modern filter-feeding, tube-dwelling annelid worms called sabellids. Since they are non-mineralized, sabelliditid tubes cannot properly be called shelly fossils.

The first mineralized fossils on the Russian Platform are the tubular shelly fossils of the Lontova Horizon (the strata immediately overlying the Rovno Horizon; Sokolov and Fedonkin 1984). Many of these resemble a sabelliditid that has become mineralized. *Platysolenites* is a simple mineralized shell that may have been originally siliceous. *Onuphionella* (figures 8.3 and 8.4) is a simple tube formed of imbricate mica flakes. Also occurring in the Lontova is *Aldanella*, a probable snail. These biomineralized shelly fossils are characteristic of the beginning of the Cambrian explosion.

Throughout the world, both biomineralization and increased burrowing were part of the animal explosion. The increase in the diversity of burrowers occurred slightly before that of shelly fossils. Together, these radiations signify a profound change in the ecology of

FIGURE 8.3. *Onuphionella durhami*, a tubular fossil formed by imbricate layers of mica flakes. Length of tube 7.5 cm. (From Signor and McMenamin 1988)

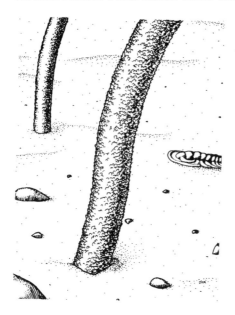

FIGURE 8.4. Reconstruction of *Onuphionella* in probable life position in the Early Cambrian sea of eastern California.

the marine biosphere. This change, however, is not solely expressed by metazoan burrowers and shells. Calcareous algae appeared at the end of the Vendian (Riding and Voronova 1982), and later radiated to become, in conjunction with the archaeocyathans, the co-creators of the first skeletal, wave-resistant reefs (Rowland 1984). The skeletons of the first calcareous algae discouraged grazers that were conditioned to consuming softer tissue. Somewhat later, the relationship between grazing and calcification acquired a curious twist. Solenopores, a group of calcareous algae that appeared in the Cambrian, are thought by Steneck (1983) to have *required* grazing to remain free of epiphytes. Solenopores underwent a sharp decline during the Jurassic (about 180 million years ago), just before their presumed descendants (coralline algae) underwent an explosive radiation. Modern coralline algae are so dependent on grazing (chiefly by parrotfish) that reproduction of some forms cannot occur without it (Steneck 1985). The descendants of Cambrian solenopores and their grazers have become mutually dependent.

CHANGES IN THE VENDIAN SEA FLOOR

AS NOTED above, abundant burrowers appear slightly before the great wave of biomineralization. Probing deposit feeders, such as the tracemaker of *Phycodes pedum*, began to excavate sediments to depths of several centimeters at the beginning of the Cambrian. Dwelling burrows several centimeters in length, such as *Skolithos*, first appeared in the Cambrian, and provided protection for filter-feeding animals. If a skeleton is broadly defined as a rigid body support, a burrow is in essence a skeleton formed of sediment (Brasier 1979). *Onuphionella* (figures 8.3 and 8.4) lived at the sea floor rather than in it (Signor and McMenamin 1988), but—in a sense—it carried the protective sediments with it, above the sea floor, by building a mica tube shell.

Movement of metazoans into the substrate had profound implications for sea floor marine ecology. One aspect of the environment that controls the number and types of organisms living in the environment is called its dimensionality (Briand and Cohen 1987). Two-dimensional (or Dimension 2) environments tend to be flat, whereas three-dimensional environments (Dimension 3) have, to a greater or lesser degree, a third dimension. This third dimension can be either in an upward or a downward direction, or a combination of both directions.

The Vendian sea floor was essentially a two-dimensional environment. The sea floor probably harbored autotrophic animals, some deposit feeders, and filter feeders. Animals surely lived in the waters above the sea floor, but there is no evidence that they were intimately linked to the sediment substrate. The flattened shapes of many members of the Ediacaran animals seem to be mirroring their environment. With the probable exception of some of the stalked frond fossils, most Vendian soft-bodied forms hugged the sea floor. Deep burrowers added a third dimension to the benthos (sea floor communities), creating a three-dimensional environment where a two-dimensional situation had prevailed. The greater the dimensionality in any given environment, the longer the food chain and the taller the trophic pyramid can be (Briand and Cohen 1987). If the appearance of abundant predators is any indication, lengthening of the food chain seems to be an important aspect of the Cambrian explosion. Changes in animal anatomy and intelligence can be linked to this lengthening of the food chain.

Most Cambrian animals are three-dimensional creatures, not flattened like many of their Vendian predecessors. Animals like mollusks and worms, even if they lack mineralized skeletons, are able to rigidify their bodies with the use of a water-filled internal skeleton called a coelom (pronounced "SEE-lum"). This fluid-filled cavity gives an animal's body stiffness, and acts much like a turgid, internal, water balloon. A coelom allows animals to burrow in sediment in ways that a flattened animal (such as, for instance, a flatworm) cannot. It is most likely that a coelom first evolved in those Vendian shallow scribble-trail makers that were contemporaries of the large soft-bodied fossils. Some of these Ediacaran burrows show evidence of peristaltic burrowing. Inefficient peristaltic burrowing can be done without a coelom, but with a coelom it becomes dramatically more effective.

SYMMETRY: EYES AND BRAINS

ANIMAL BODY plans can be lumped into two categories: radially symmetric and bilaterally symmetric. Cnidarians and echinoderms are well-known examples of radial body plans. Radial animals are symmetric about a central axis, and may have three-fold radial symmetry (as in *Tribrachidium*; figure 2.4), fourfold (as in a jellyfish) or fivefold (as in the five arms of a starfish). Radial animals may also be tubular, as in sea cucumbers and predatory priapulid worms. First known from the Burgess Shale (Conway Morris 1977), priapulids are named for Priapus, the Greek god of fertility, typically represented with "a regal and ready copulatory organ" (Pearse et al. 1987:447). The pentagons of spines surrounding the mouth area of a priapulid worm indicate five-fold radial (pentameral) symmetry, with the axis of symmetry running the length of its elongate body. Sea cucumbers or holothurians, a mostly soft-bodied group of modern echinoderms, are another example of an animal with an elongate body and pentameral symmetry. Most tubular animals, however, are bilaterally symmetric.

You are a good example of a bilateral animal with imperfect bilateral symmetry. Most people have two eyes, two ears, two lungs, and two kidneys, and their internal bilateral symmetry is broken only in details such as the displacement of certain internal organs such as the heart and liver to one side of the body or the other.

137

Annelid worms are also bilaterally symmetric. Annelids or annelid ancestors were the most likely fabricators of those Vendian peristaltic burrows, so the title of "first coelomate animal" probably belongs to a Vendian annelid. It is also probable that these first coelom-bearers were bilaterally symmetric. Bilateral symmetry is important when considering the behavior of these early coelomate animals. The most likely animal to evolve a brain is one with bilateral symmetry.

Concomitant with the emergence of animals during the Vendian was the origin of brains. The Cambrian explosion was the first cerebralization or encephalization event. As part of the increase in the length of the food chain discussed above, higher-level consumers such as top or keystone predators established a mode of life that requires the seeking out and attacking of prey. These activities are greatly aided by having a brain able to organize and control complex behavior. The bilateral Vendian burrowers may have had some degree of cerebralization. Although we might have had a hard time telling one end of a Vendian worm from the other (had it left a body fossil), the worm must have known its posterior end from the burrow it was excavating. But there is no evidence that animals were particulary "smart" until the advent of complex burrows and fossil predators such as *Protohertzina* (figure 4.6).

The complex shapes of some Vendian trace fossils (Fedonkin 1978) imply a moderate amount of programmed behavior. Fedonkin (1978) shows how simple traces such as a sinusoidal burrow or spiral track can be joined to more complex and potentially more efficient combinations of deposit feeding behavior. Most of the trace fossil types known appear in the Vendian and Cambrian, strongly suggesting that brains capable of combining two or more simple burrowing programs had appeared at that time. This is direct evidence of Calvin's (1986) argument that the appearance of brains institutionalized rapid innovation. Assuming *Protohertzina* did closely resemble chetognaths (arrowworms), then the dorsal nerve ganglion of a chetognath is representative of the type of brain one might expect to find in the earliest bilateral metazoan predators.

Modern arrowworms can give an indication of how primitive predators live and locate prey. Chetognaths are almost wholly planktonic, but this does not mean that chetognaths and chetognath-like animals could not have had an influence on benthic biotas. *Spadella cephaloptera* is a typical chetognath except that it has taken up a benthic existence. About 1 cm in length, it attaches itself to the sea

138

floor by means of adhesive cells along the underside of its tail. It deftly snatches prey (usually small arthropods) without having to move from its attachment site (Pearse et al. 1987). At its head end, *Spadella* has grasping spines, paired clusters of simple eyes, and a ciliated loop (characteristic of chetognaths) thought to provide a sensory function. The eyes of chetognaths are simple pigment cups and are not image forming; planktonic chetognaths detect their prey by means of bunches of sensory cilia distributed along the length of the body. Thus chetognaths exist at a relatively low degree of cerebralization; indeed, a second "brain" (the ventral ganglion) is present below and behind the head-end ganglion, and may help process information received from the ventral sensory hairs.

Specialized light receptors seem to be a characteristic of all animals and many other types of organisms; Salvini-Plawen and Mayr (1977) have shown that photoreceptors have originated independently in at least forty and perhaps as many as sixty groups. Most animal phyla have at a minimum several pigmented eye spots. But advanced vision (i. e., compound or image-forming eyes) tied directly into a centralized brain is not common or well developed until the Cambrian. The tendency to have eyes is more pronounced for bilateral than for radial animals.

Trilobites, like modern crustaceans such as lobsters, probably had a nervous system formed of a string of paired ganglia running along the ventral side of the body. Chances are, however, that the anteriormost ganglion in trilobites was enlarged and fused into a brain, and processed the information received from the animal's eyes, antennae, and other sense organs, as well as controlled the stomach that was ensconced in the center of the cephalon. As mentioned in chapter 7, some of the earliest trilobites had large compound eyes. Trilobites were probably not particularly smart by modern standards, but chances are that their behavioral capabilities far outstripped any that had existed during the early Vendian.

Despite the preceding discussion, braininess is not a requirement for being a successful predator. Many modern cnidarians are predatory. Among the higher metazoa, sea spiders, or pycnogonids, are successful predators on a variety of sessile marine animals. Sea spiders have a primitive body plan that may date back to the Cambrian (although fossils are only known from Devonian strata). The brain and head of a sea spider is so weakly developed that it is difficult to tell its head end from its hindquarters, and early descriptions of the Devonian fossils confused the two (Pearse et al. 1987).

Centralized sensory reception and brains, however, are highly advantageous for predators that seek out mobile prey. *Anomalocaris* (figure 7.1), like early trilobites, had large eyes. Actively moving or vagile predators are, as a rule, smarter than their prey, because of the more rigorous requirements of information processing in a predatory life mode. *Anomalocaris* as a seek-and-destroy top predator may have been the brainiest Early Cambrian animal. Modern cephalopods such as squids and octopi are nearly all predators and have well developed image-forming or "camera" eyes that are very similar to the eyes of vertebrates. The first fossil cephalopod shells (Chen and Qi 1981) appear in the Late Cambrian, indirect evidence that an advanced pair of image forming eyes connected to a brain had evolved by then in these bilateral creatures. Weakly skeletonized "protonautiloid" cephalopods may have been present even earlier in the Cambrian (Hutchison 1961).

But what of the intelligence of radial metazoa such as cnidarians and echinoderms? It would be difficult to prove that a crinoid is much "smarter" than any of the sessile soft-bodied organisms of the Vendian. Skeletonization of echinoderms, however, can be viewed as a protective measure, an indirect effect of the emergence of smart predators that can actively detect and move toward prey. Passive suspension feeders need to be out in the open where the food is, and as such are at great risk of being attacked by predators (Vermeij 1987). Recall the new Ediacaran soft-bodied impression, *Arkarua* (figure 4.17). *Arkarua* has pentameral symmetry, and Gehling (1987) interprets this fossil as the earliest echinoderm. Secreting a skeleton is perhaps the easiest way for a soft-bodied organism such as *Arkarua* to defend itself against predators. Indeed, disc-shaped Cambrian edrioasteroids are very similar to *Arkarua*, and *Arkarua* deserves further study to determine whether or not it really is the earliest known echinoderm. The filter feeding grooves in edrioasteroids are covered by interlocking, calcite plates. Does an edrioasteroid skeleton represent an attempt to protect a passive, Garden of Ediacara lifestyle?

To ask another question, why didn't brains and advanced predation develop much earlier that they did? A simple, thought experiment may help address this problem. Consider a jellyfish 1 mm in length and a cylindrical worm 1 mm in length. Increase the size (linear dimension) of each (by growth of the individual or by evolutionary change over thousands of generations) one hundred times. As noted in chapter 2, area-volume effects come into play. The worm will need internal plumbing because of its cylindrical body. The jellyfish won't be as dependent on plumbing because its body has a higher surface area. The nervous system of the worm, concentrated at the head end in the ganglion or brain, is also affected by the surface-volume considerations. Our enlarged, 10 cm long worm will possess a brain which has a volume one million times greater than the brain of its 1 mm predecessor (assuming that the shape of the brain remains constant). The jellyfish will also get more nerve tissue as it enlarges. But its nervous system is spread out in a netlike fashion; at most, its nerve tissue will be concentrated at a few radially symmetric points. The potential for complex and easily reprogrammed behavior, as well as sophisticated processing of sensory input data, is much greater in the animal with the million times larger brain (containing at least a million times as many brain cells as its tiny predecessor). Complex neural pathways are more likely to form in the larger brain. This implies no mysterious tendency for animals to grow larger brains; perfectly successful, advanced animals (echinoderms) and even slow-moving predators (sea spiders) get along fine without much brain. But centralized nerve tissue can process information better than a nerve net and control more complex responses to stimuli. Once brains were used to locate food, the world would never again be the same. This can be thought of as a "brain revolution" that permanently changed the world a half billion years ago. The brain revolution must have influenced the skeletonization event of Cambrian animals, because, as Hutchison (1961) astutely noted, the thickness of Cambrian skeletons goes above and beyond the requirements for mere muscle attachment.

Shells influenced the evolution of new body plans. The stalked echinoderms (sometimes called pelmatozoans) used skeletons to great advantage by forming a shelly stem that allowed them to extend high into the water column. This development increased the dimensionality of Cambrian benthic environments in an upward direction. The elongate, stalked body type proved to be extremely successful later in the Paleozoic, when crinoid and blastoid pelmatozoans became abundant. Brachiopods may have first evolved shells as a protective measure. Shells in articulate brachiopods, in addition to protecting the delicate filter feeding apparatus call the lophophore, enable the brachiopod to direct a precise stream of laminar flow through the lophophore for efficient filter feeding (LaBarbera 1981). The evolution of shells for defense against predators thus opened the way for improved types of passive filter feeding.

141

The last two examples (anchored suspension feeding above the sea floor and directed-flow filter feeding) show that the evolution of skeletons opened new niches that had never before been available until the Cambrian. So many different types of skeletonization and new life habits arose during the Cambrian, however, that this event cannot be solely characterized as a breakthrough in animal biomineralization or even as a result of evolution of the brainy predators that initiated the predator-prey escalation seen throughout the rest of geologic time (Vermeij 1987). Ecological feedback certainly seems to be involved in the biotic events of the Vendian-Cambrian transition. One wonders, however, why the initiation of this feedback loop was delayed millions of years until just before the beginning of the Cambrian. Were there environmental changes that could have led to these ecological changes?

Changes occurred in the nature of the sea floor during the Vendian-Cambrian transition. It has been suggested that Vendian monerans carpeted the sea floor, forming a resistant cyanobacterial scum over large stretches of sediment surface (Seilacher et al. 1985), even in marine environments where today, current action and burrowing activity usually stir up the sediments. This moneran mantle is preserved as folds, ruffles, and tears in sediment layers that were stabilized by binding algal mats. Evidence for moneran mantling occurs in direct association with Ediacaran soft-bodied forms (Gehling 1986, 1987).

In the Rawnsley Quartzite of South Australia, soft-bodied fossils are found on the bottom of sandstone beds. The sand was deposited by periodic storm events when sand was carried to deep, undisturbed waters. Gehling (1986) suggests that soft-bodied Ediacaran creatures were able to preserve as fossils, despite being buried by massive sands, because the sediment on which they rested was coated by a moneran film that made the substrate resistant to erosion by the current surges associated with the deposition of the storm sands. Gehling (1987) argues that colonization of the seafloor by moneran mats below the depth of normal wave disturbance was possible, due to the lack of grazing and predatory organisms (the mat, of course, would have to reestablish itself after each storm sand deposit). Also, he suggests that the absence of surface grazing and infaunal burrowing animals allowed the preservation of gas-induced convolutions in the sediment (Gehling 1987).

Recall *"Brooksella" canyonensis*, the enigmatic billion-year-old Grand Canyon object discussed in chapter 3. We propose that the gas bubble interpretation of *"Brooksella" canyonensis* is correct, except that the gas escape event formed a series of pockets under a Vendian-style moneran mat. Cloud (1968:27) is essentially correct in suggesting that *"Brooksella" canyonensis* is a preserved gas blister, but blisters like this one may not be able to form today, because moneran mats of this character no longer form in subtidal environments. Further evidence for this interpretation can be seen in the Clemente Formation pit and mound structures (figure 3.4). Note the small blisters on the flanks of one of the sand volcanos. We interpret these blisters as representing the failed attempt of a column of fluidized sediment to make its way through a moneran mat to the sediment surface. The fluidized sediment was impeded on its flow path upward, and it accumulated to form a vent-clogging plug of sediment. Plugs such as these can occur in sand volcanos formed in sediment that is not bound by moneran mats, but finding two such plugs so closely spaced is rare, and suggests that more than just gravity was impeding the upward flow of the fluidized sediment. We interpret these closely spaced sediment plugs (which might more properly be referred to as sediment-filled blisters) as indirect evidence for microbial binding of the Precambrian sea floor surface that eventually became part of the Clemente Formation.

There is direct evidence for the organisms that constructed these mats. The lower Vendian Chichkanskaya Formation of the USSR has flat-pebble conglomerates that contain well-preserved, non-calcified filamentous and spheroidal moneran microfossils (Ogurtsova and Sergeyev 1987). Knoll (1982) described the moneran paleoecology of silicified mat fragments preserved in the Upper Riphean (800 to 700 million years ago) Draken Conglomerate of Spitsbergen (a large island due north of Scandinavia). The Draken mats formed in a protected limey lagoonal environment. As in the Rawnsley Quartzite discussed above, the Draken Conglomerate shows evidence of storm-influenced deposition. Periodic tempests ripped up the thin mats and redeposited them, forming flake conglomerates, the sedimentological equivalent of dandruff. These flake conglomerates resemble flat pebble conglomerates, except that the individual Draken mat shards are usually thinner and more flexible than the flat pebbles. Knoll (1982, 1985) shows that the mat is composed of the tubular sheathes of filamenteous monerans, tightly interwoven to form a resistant feltlike layer. *Eomycetopsis* (figure 8.5) is an important mat-forming moneran in Riphean and Vendian moneran mats.

Microbial mats such as these are largely (but not exclusively) restricted today to environments unfavorable to the metazoans that would otherwise feed on and disrupt them. Knoll (1985) notes that the structure and microbial make-up of the Draken mats are comparable to what one sees in Recent mats growing in similar environments. It seems reasonable, then, to infer that moneran mats in the Precambrian were similar to later mats except for their being much more widespread. The mat-forming organisms of the Rawnsley Quartzite probably harbored filamentous monerans similar to *Eomycetopsis*.

As noted earlier, mat builders may have tried to defend themselves against grazing metazoa by forming thicker microbial sheathes, sometimes strengthened by calcium carbonate. The spaghettilike *Girvanella* is abundant in, but not restricted to, the Cambrian (Rezak 1957), and has a robust calcium carbonate sheath.

Another way for microorganisms to avoid grazers is to move into the sediments. Endolithic ("within the rock") microorganisms bore into calcium carbonate sediments and shells, and are involved in the mechanical destruction of both. Figure 8.6 shows the phosphatic casts of endolithic microorganism borings in an echinoderm skeletal plate. The endolithic microbes bored into the shell, and the boring tunnels were subsequently filled with phosphatic precipitate to form casts of the borings. When we recovered this fossil from an acid bath, the outer part of the echinoderm plate was dissolved, exposing the

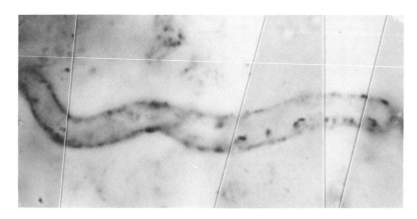

FIGURE 8.5. *Eomycetopsis*, a filamentous moneran, was an important mat building organism in many Precambrian fossil localities. This specimen is from the Kheinjua Formation in central India. Filament diameter 5 microns. (From D. McMenamin et al. 1983)

144

boring casts which originally were within the shell. Such endolith casts are frequently seen on Early Cambrian shelly fossils (Runnegar and Bentley 1983, their figure 6c). Although the rise of animals decreased the stability of microbial mats, the appearance of shells provided new substrates for endolithic microorganisms.

In addition to possibly causing the decline of stromatolites and the appearance of skeletonized microorganisms, the appearance of metazoan grazers seems to have destabilized the microbial mat, allowing the accumulation of sand-sized calcium carbonate fecal pellets and resulting in the deposition of flat-pebble conglomerates (figure 8.7) during the Cambrian (Sepkoski 1982). Flat-pebble conglomerates (very similar to the flake conglomerates discussed above) are distinctive sedimentary rocks composed of flat chips or disks (often 5 to 10 cm in greatest dimension) set in a matrix of finer silty limestone or clay. This type of sedimentary rock is uncommon

FIGURE 8.6. Phosphatic casts of endolithic microorganisms on a degraded fragment of Lower Cambrian shell (possibly an echinoderm plate). From the Buelna Formation, Sonora, Mexico; scale bar = 0.1 mm.

before the late Vendian and after the Ordovician, but during the Cambrian and the Vendian-Cambrian transition it was frequently formed, particularly in western North America. Sepkoski (1982) argues that after grazing became common enough to prevent abundant stromatolite formation, the limey, exposed sea floor underwent rapid surface lithification. This fragile, limey crust was then broken up and transported by storm events, resulting in the deposition of a flat-pebble conglomerate bed. When the amount of infaunal burrowing reached the high levels observed during the Late Cambrian and Ordovician (Droser and Bottjer 1988), even the calcium carbonate crusts (and the flat pebble conglomerates derived from them) were prevented from forming because of the intense sediment churning activities of burrowing animals.

Burrowers and grazers created new environments for themselves and other metazoans by changing the substrate of the sea so that it

FIGURE 8.7. Flat pebble conglomerates are only common in the late Vendian and the Cambrian, after stromatolites became rare and before infaunal burrowing became widespread. The example shown here is from the Lower Cambrian Puerto Blanco Formation, Cerro Rajón area, Sonora, Mexico; scale bar = 10 cm.

was no longer favorable for the growth of stromatolites, moneran scums, or flat-pebble precursor crusts. The increasing prevalence of burrowing also had a direct influence on skeletonization. Fedonkin (1987) notes that skeletonization allowed sessile forms to develop a lifestyle that was less easily disturbed by the soft-bodied burrowers.

In an almost allegorical sense, the flat-pebble conglomerates represent the breaking up and destruction of the two-dimensional Garden of Ediacara. C. W. Thayer (1979) has used the term "biological bulldozers" to refer to the activities of ploughing deposit feeders. This type of bulldozing, in which a coelomate organism is displacing substantial amounts of sediments as it gathers food, is first known from the late Vendian to Early Cambrian, and indeed forms part of the trace fossil explosion. The Garden of Ediacara, with a sea floor dominated by monerans, appears to have been literally bulldozed out of existence.

The shells of metazoans caused further problems for the mat-forming monerans. Monerans have difficulty trapping and binding coarse bioclastic sediment (Awramik and Riding 1988), and the change in the texture of the sea floor sediment when shells and fragments of shells became abundant surely must have had a detrimental affect on the ability of monerans to form microbial mats and to stabilize marine sediments. The shelly grit formed by discarded shells may have had an even worse effect on the formation of moneran mats than overgrazing of moneran communities by the metazoans themselves.

There is one counter-example in which the Cambrian metazoans seem to have formed a cooperative complex with their moneran neighbors, rather than destroying moneran habitat. Thrombolites are organically-formed sedimentary structures that are very similar to stromatolites, except that instead of the laminar fabric of stromatolites, they have a clotted internal structure. There is evidence that thrombolites were moneran/algal/metazoan communities; in other words, they were stromatolites riddled and colonized by metazoans and partly formed by calcified microorganisms (Kennard and James 1986). Like flat-pebble conglomerates, thrombolites are essentially a Cambrian to Lower Ordovician phenomenon, and decline in abundance sharply after the end of the Cambrian. Apparently, the composite communities that formed thrombolites did not prove to be long-term successes like the stromatolites before them, possibly because of the destabilizing effects of burrowing activity that escalated throughout the lower Paleozoic (Droser and Bottjer 1988).

Thus burrowing seems to have restructured the sea bed to the advantage of burrowers, an example of the positive feedback that seems to have been such an important component of the Cambrian explosion. The dimensionality of the sea floor increased in a downward direction, as burrowers excavated to successively greater depths. As well as providing more area for excavation by deposit feeding metazoans, the disturbance of the moneran scum/sea floor crusts also allowed buried nutrients and organic carbon to reenter the water column (Lambert et al. 1987).

MARINE CHEMISTRY OF MIROVIA

IN ADDITION to increases in the ecological dimensionality of the sea floor (in upward and downward directions), marine chemistry and nutrient levels underwent significant changes during the Vendian-Cambrian transition. Brasier (1986) and Conway Morris (1987a) discuss the geochemical clues indicating major oceanic chemistry perturbations that may have had a direct bearing on the biological evolution near the Vendian-Cambrian boundary. Marine sulphur is a participant in these fluctuations. Measurements of the ratio between light (^{32}S) and heavy (^{34}S) isotopes of sulphur can be made from evaporite minerals. Evaporites are sediments deposited as a result of supersaturation of sea water due to evaporation. They include minerals such as halite (rock salt), anhydrite (pure calcium sulfate) and gypsum (alabaster). Measurements show that the seawater ratios of heavy to light sulphur have fluctuated widely since the Proterozoic. Holser (1977) recognizes three large "spikes" or sudden changes in the isotope ratio data of the last 700 million years. At the times of these "spikes" or excursions, evaporites formed which were enriched in the heavy isotope of sulphur. The largest heavy sulphur "spike" is called the Yudomski event, named for late Proterozoic sediments of the Siberian platform. The Yudomski event occurred very close to the Vendian-Cambrian boundary (Conway Morris 1987a).

The amount of heavy sulphur in the sea is balanced by two processes. The first is the precipitation of sulphur by monerans in anoxic (oxygen-free) sediments. These monerans (similar to the ones living in the guts of the vent fauna autotrophic animals) take hydrogen sulfide from marine sediments and use it as an energy source. These microbes combine the hydrogen sulfide with iron in the sediments to form sulfides such as the mineral pyrite, an iron sulfide. When the monerans do this, they preferentially use hydrogen sulfide composed of the lighter isotope of sulphur. This fractionation of isotopes is analogous to a bricklayer who tries to keep his or her arms from tiring so quickly by preferentially using a lighter variety of brick. Both the microbes and the bricklayer are trying to minimize their energy losses (or in other words, conserve energy), and one way to do this is to use less massive building materials (either lighter isotopes or lighter bricks).

The net effect of bacterial fractionation of sulphur isotopes is that, as the pyrite-laced sediments of the sea floor—where the monerans live—accumulate and become enriched in light sulphur, the sea water itself becomes depleted in the lighter isotope. Another way to put it is that sea water becomes enriched in the heavier isotope, and this increase can be reflected in the sulphur isotopic signature of minerals such as evaporites that are inorganically precipitated from sea water. Evaporites of the Yudomski event are severely depleted in light sulphur, indicating that the ocean of the time (Mirovia) was depleted in this isotope as well. This must in some way reflect an episode of enhanced production of anoxic sulphide by monerans.

The second process influencing the balance of sulphur isotopes in the ocean is oxidation of sulphides. Pyrite is an unstable mineral when exposed to oxygen in the atmosphere at the earth's surface. It has been unstable at the earth's surface for approximately two billion years, or ever since the earth's atmosphere became oxygen-rich. Pyrite exposed at the surface ever since then oxidizes to form iron oxide minerals and hydrogen sulfide.

Where did the light isotope of sulphur hide during the Vendian-Cambrian transition? Evaporites commonly form in restricted marine basins that have limited access to the open ocean and are thus liable to dry up. Curiously, the formation of a brine in a marine basin can promote anoxic conditions in the sediments below. Conway Morris (1987a) suggests the following scenario for the Yudomski event.

As Rodinia began to fragment as a result of rifting, narrow marine basins with poor circulation formed between the parting continental fragments. These narrow basins were ideal sites for both deposition of anoxic bottom sediments (enriched in light sulphur because of moneran activity) *and* for excess evaporation of sea water. Such evaporation causes brines to form, and if this evaporation is carried to extremes, evaporite mineral deposits will precipitate on the bottom of the narrow rift ocean. These evaporites, since they reflect the chemistry of the waters from which they are precipitated, will be enriched in heavy sulphur and will serve as reservoirs of heavy sulphur until they are dissolved in water again. As the rift basin oceans increased in width as a result of continued continental drift, open marine conditions with good circulation would eventually become established in the new oceans and evaporites would cease to form.

Concurrent with this widening of rift oceans was a global rise in sea level that drowned the edges of all continents during the Cambrian. The spreading centers of the new oceans, along the axes of the elongate trough seas, were sites of creation of new, hot oceanic crust. As this new ocean crust formed, there was a corresponding destruction (by subduction) of ancient, cold, and less bouyant Mirovian sea floor crust. The exchange of new, hot sea floor crust for old and cold Mirovian crust caused the displacement of great quantities of water onto the continents, apparently causing the Cambrian transgression (global rise in sea level). An incidental result of this transgression was the destruction by dissolution of the evaporites (enriched in heavy sulphur) by the rising sea waters. This heavy sulphur was suddenly injected into the sea, resulting in the Yudomski event.

It appears that both the buildup of heavy sulphur in the evaporites and the sudden release (of sulphur and other, more nutritious, chemicals) as a result of transgression-induced dissolution influenced the Precambrian-Cambrian transition biotas in ways that encouraged the metazoan explosion. This transgression also provides a partial vindication and explanation for Walcott's (1910; see chapter 1) discredited Lipalian interval. During a transgression, one of the first marine environments to encroach on the shoreline is the nearshore zone of breaking waves. This breaker zone is very erosive, and as the transgression proceeds, the shoreline with its erosive wave action moves progressively further and further inland. The erosive environments that occurred during the early stages of the Cambrian transgression are unlikely to have left a continuous record of deposition; therefore, many Vendian-Cambrian transition sedimentary sequences have an unconformity exactly at the Vendian-Cambrian boundary. The transgression wave front often cut down into Vendian sediments as it passed through any particular site, further magnifying the gap in the sedimentary record at this boundary. Unfortunately for correlation, the age of the Vendian-Cambrian gap is not the same in widely separated stratigraphic sections because of the nonsynchronous nature of the transgression and because of the varying amount of erosion suffered by stranded Vendian sediments during the transgression.

Areas further inland experienced the front wave of the transgression later than areas more seaward. In Sonora, lowermost Cambrian sandstones begin the Cambrian sequence (M. McMenamin et al. 1983), whereas at some sites in the continental interior (such as South Dakota; Lochman-Balk 1964), the first Cambrian sediments were deposited as Upper Cambrian sandstones. Therefore, the base of Cambrian sedimentation gets younger as one goes further inland, as a simple result of the fact that it took millions of years for the waters of the Cambrian transgression to finish their inundation of the North American continent.

At any particular site, sandstones are usually the first sediments deposited after the erosive initial part of the transgression passed through. Unfortunately, such sandstones are often poorly fossiliferous, and the fossils present are of much fewer types of organisms than are usually encountered in deeper water sediments. This makes most Vendian-Cambrian boundary sites doubly unpromising for placement of a "golden spike" stratotype boundary. Not only is there discontinuous sedimentation near the boundary, but fossils are likely to be rare in the coarse sandy sediments near the boundary that did manage to avoid being eroded. The fossil record improves at any site as the sandstones of the initial transgressive phase pass by and deeper water sediments, such as shales and limestones with richer faunas, begin to accumulate.

There are a few sites (in South China, Siberia, Newfoundland) where the magnitude of the Vendian-Cambrian gap seems to be minimal, and thus the Lipalian concept in its extreme form (i.e., no sedimentation during the boundary) is untenable. But one aspect of the Lipalian interval must be retained. That is the fact that similar continental margins were subjected to the erosive effects of a transgression at more or less the same time. This transgressive episode was occurring all over the world at a time when most continental margin areas were undergoing similar tectonic events, namely, uplift, rifting, and subsidence resulting from the breakup of Rodinia. Not only were the evaporites remobilized into sea water by the rising waters, but sediments and nutrients that were ponded in lagoons and nearshore terrestrial environments during the late Vendian were also brought into deeper water. Times of relatively stable sea level can result in the ponding of sediments in lagoons and lowland areas near the sea (Mount and Rowland 1981), and these sediments can get moved into the sea when sea level fluctuates.

The ponding effect would bring sediments derived from continental erosion into contact with sea water. Keto and Jacobsen (1985) have calculated the variations in the ratio of heavy strontium (^{87}Sr) to light strontium (^{86}Sr) in seawater of the past 750 million years, and conclude that the highest value in the mass flux ratio occurred during the Vendian-Cambrian transition. This peak corresponds to maximum exposure of sea water to the products of continental weathering and erosion (Keto and Jacobsen 1985).

The strontium peak is possibly a result of two factors. The first factor would be increased erosion rates at the margins of the incipient rift seas, where high heat flow from the earth's interior caused uplift (and ensuing erosion) of areas of continental crust adjacent to the rift. This rift-induced swelling in the continental crust gets uplifted in relation to the surrounding continent, which leads to the ironic situation of having an uplifted region (rift margin) immediately adjacent to a downdropped region (rift basin; Stewart and Suczec 1977; figure 8.8).

The second factor would be the "de-ponding" effect resulting from the transgression discussed above. Rising seas would bring these continental sediments in contact with open ocean waters. A test could be made of clastic sediments deposited near the Vendian-Cambrian transition interval to determine whether the proportions of continental sediments in marine waters increased significantly at that time.

Many nutrients, such as phosphorous and nitrogen, are limiting to the biota in marine environments. In other words, the number of organisms in a given environment—such as the open ocean surface waters—are limited by the availability of these nutrients. Four factors seem to have worked in concert to increase the amounts of nutrients in marine waters at the Vendian-Cambrian transition. First and second, discussed above, were the sediment-stirring activities of actively burrowing metazoans and the erosive waves of rising sea level. The third factor was the increased erosion off of continent margins uplifted by rifting. The fourth involved a major change in paleoceanographic conditions.

As Rodinia was sundered into numerous fragments at the Vendian-Cambrian transition, Mirovia shrank to make room for the widening rift/ocean basins appearing between the continental fragments. This undoubtedly resulted in substantial changes in global sea floor bathymetry. As noted earlier, the buoyancy of the newly rifted ocean crust caused the Vendian-Cambrian transgression. As circulation between these widening oceans and what remained of Mirovia improved, deep, nutrient-rich waters were able to move close to shore in the rift oceans. In narrow seaways and other situations where oceanographic conditions are favorable, deep ocean water is brought to the surface in a process called upwelling (Cook and McElhinny 1979). Sites of upwelling in modern oceans are associated

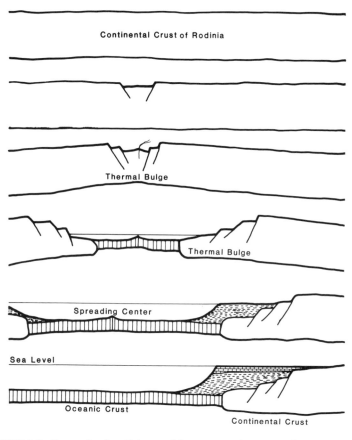

Continental Crust of Rodinia

Thermal Bulge

Thermal Bulge

Spreading Center

Sea Level

Oceanic Crust

Continental Crust

FIGURE 8.8. Stages in the rifting and formation of new sea floor as a result of the breakup of Rodinia at the end of the Precambrian. The sequence proceeds from top to bottom. Note the uplift due to a thermal bulge during the incipient stages of rifting. (Modified after Stewart and Suzcek 1976)

153

with extremely high organic productivity, major fisheries, and deposition of phosphate-rich sediments on the sea floor below.

Abundant and economically important phosphate deposits called phosphorites are known from Vendian and Cambrian sediments throughout the world. Cook and Shergold (1984) hypothesize that a major increase in the phosphorus content of shallow marine waters occurred during the Vendian-Cambrian transition. They see this increase as a consequence of enhanced oceanic upwelling, following a period of decreased overturn (Cook and Shergold 1984). In addition to resulting in the abundant phosphate deposits of this age, Cook and Shergold (1984) postulate that this increase in nutrients (eutrophication) drove the Cambrian explosion by increasing the availability of phosphate in marine waters and making it easier to form phosphatic skeletons.

A stumbling block to this hypothesis is that calcium carbonate shells seem to be at least as prevelant as phosphatic shells at the Vendian-Cambrian boundary. It may be more proper, then, to see the episode of phosphate deposition as part of a larger event in which the oceans saw a sudden increase in the amount of available nutrients. The Vendian-Cambrian phosphorites may have formed as a result of paleoceanographic factors favoring upwelling on an unprecedented scale (Cook and Shergold 1984; Kidder and Swett 1989), at a time when many continents were at, or near, the equator.

Warm equatorial waters and abundant nutrients are, in biological terms, a potentially explosive combination. But can this really explain the profound biological turnover that occurred at the Vendian-Cambrian transition, when new body plans and new ways of living emerged in concert? This turnover can be viewed as a restructuring of the ecology of the marine biota, perhaps as an direct result of changes in nutrient supplies in the shallow waters of the newly formed rift seas forming between the remains of Rodinia. Although it is not yet possible to prove a link between marine nutrient supplies and the feeding strategies in marine animals of the Vendian-Cambrian transition, it is possible that an increase in nutrient supplies spurred the rapid appearance of predators at the base of the Cambrian (but at the top of the food chain!). Hallock (1985) notes that areas of upwelling favor heterotrophic animals, and she has results (Hallock 1981) showing that the energetics of the animal host-photosymbiont association are most favorable in low nutrient waters. By supplying abundant nutrients (including phosphorous), the Vendian-Cambrian episode of marine eutrophication would have favored hard path animals over photosynthesizing soft path animals.

Several factors working in concert, some of them self-reinforcing, may have made conditions less and less favorable for soft-bodied Ediacaran creatures. Positive feedback loops involving diverse processes such as stromatolite grazing and upwelling led to a massive upheaval of the benthic marine ecosystem.

The late Vendian extinctions and the dominance of bacterial sheathes and *Bavlinella* in the acritarch residues of this time can be explained as part of this marine ecological restructuring. Recall that during the second half of the Vendian, acritarch and metazoan diversity plummeted, and only later exploded (animals in the latest Vendian) or recovered to previous levels (acritarchs in the Early Cambrian). The onset of the transgression and the advent of abundant burrowers, plus the increased erosion rates off of the uplifted continental margins of the rift seas, probably led to increases in the amount of suspended sediments in coastal marine waters. This began sometime in the late Vendian, when rifting of Rodinia was well underway.

Interestingly, resuspension of marine sediments has recently been shown to stimulate the growth of heterotrophic microplankton (Wainright 1987). Both bacteria (monerans) and protozoans (protists such as foraminifera) benefit greatly from suspension of sediments. Sediment particles in suspension become more nutritious the longer they remain entrained in the water column, because they become more densely packed with bacteria. These bacteria act as a food source that stimulates the growth of heterotrophic microbes, and Wainright (1987) notes that this increased production is an important source of high-quality food resource for consumers higher in the trophic pyramid. In other words, stirring up of sediment has an important stimulating effect on smaller hard path organisms, which in turn are fed upon by animals higher up in the trophic hierarchy. In modern environments, level-bottom marine environments consisting of quartz and clay-rich sediments frequently have a centimeter-thick (or thicker) mantle of easily resuspended mud that results from physical reworking by deposit-feeding animals and continuous settling-out of silt and clay from ambient wave and current action (Bokuniewicz and Gordon 1980). The particles forming this soupy substrate, or rather the monerans growing on the particles, constitute a rich food source for filter feeders.

Although destabilized sediment is a better food source, it can cause difficulties for organisms adapted to firm substrates and water. Very soupy substrates force sea-floor filter feeders to grow upwards into the water column or face strangulation by having their filter-feeding apparatus clogged by the turbid water at the sediment-water interface. On firm substrates, simple cylindrical metazoan burrows are unlined and can be indistinct as fossils. In mushy substrates, burrowers must firm up their burrows by lining them with mucous. When fossilized, lined burrows have very distinct outlines (Rhoads 1970). Most Precambrian cylindrical burrows are unlined and indistinct, indicating that the sediments in which they formed were firm, not soupy.

Loss of sediment stability on the sea floor had profound implications for the late Vendian marine world. Rising sea level and the flooding of continental areas resulted in the creation of more shallow marine sea floor and more storm-influenced deposition. Severe storms cause the deposition of thin, often sandy, storm sediment sheets that can spread significant distances offshore. These desposits are known by the picturesque name tempestites, and tempestites are very commonly found in uppermost Vendian and Lower Cambrian sequences (Mount and Rowland 1981). Flat-pebble conglomerates (Sepkoski 1982) are one type of tempestite common to this interval of geologic time (see figure 8.7). One undoubted result of the formation of a storm deposit is resuspension of marine sediments. In addition to acting as a floating template for the growth of bacterial food sources used by filter-feeding heterotrophs, the suspended sediments encourage hard path feeding strategies at the expense of soft path strategies. Photosymbiosis, for example, requires clear water capable of transmitting usable light to some depth.

As mentioned in chapter 6, supercontinents such as Rodinia have a harshening effect on global climate, and the presence of the supercontinent was certainly a factor increasing the severity and extreme seasonality experienced during the Varangian glaciation. Valentine and Moores (1972) have suggested that large seasonal environmental extremes during the Varangian glaciation were responsible for the Vendian delay in the metazoan explosion. Conditions during the late Vendian (Rodinia breakup, transgression, overgrazing of stromatolites and algal mats, storms across shallow seas) encouraged the stirring up of sediment and the growth of bacteria in the water column, a major step toward creating a stable base to the food chain which heterotrophic organisms could exploit. Recall that this "clouding of the waters" could have a negative effect on some autotrophic organisms.

The same low nutrient (oligotrophic) waters of McMurdo Sound, Antarctica, that harbor the direct nutrient-absorbing foraminifera discussed in chapter 7 (Delaca et al. 1981) have yielded new information concerning the feeding of echinoderm larvae. Rivkin et al. (1986) show that larvae of the antarctic echinoderm *Porania antarctica* preferentially ingest bacteria and avoid eating photosynthetic phytoplankton. A difficulty of the Antarctic environment is that production of photosynthetic plankton is severly limited by the severe climate and short growing season. Seasonality in a cold-dominated environment such as this one has a major influence on resident organisms. The Antarctic bacterial food supply is much less seasonal than the "boom-and-bust" of phytoplankton populations (such as dinoflagellates), and it is no wonder that the larvae of *Porania antarctica* prefer the more dependable food source.

The conclusions of this chapter can be summarized in four relatively simple diagrams. Figure 8.9 is a spectrum of feeding types, ranging from photoautotrophy at the "soft" end of the scale to the keystone predator (or parasites of a keystone predator) at the extreme "hard" end of the scale. Figure 8.10 compares the dominant trophic strategies, animal body plans, and animal physiologies of Garden of Ediacara and Heterotrophic Pyramid ecosystems. In heterotroph-dominated trophic pyramids, it is not only the higher-level carnivores that exerted important controls on the nature of the Cambrian ecosystem. Grazers feeding on stromatolites and algal mats cleared the sea floor for the heterotroph-dominated communities, and one could argue that if metazoan grazers were to be removed from the face of the earth, that stromatolites and algal mats would return to the sea floor in all their pre-Vendian glory. The algal "lawn" may require constant "mowing" to allow a marine ecology as we know it.

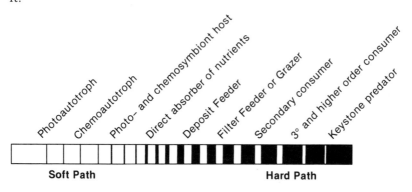

FIGURE 8.9. Spectrum of feeding types in generalized ecosystems, showing the "soft path" and "hard path" ends of the spectrum.

Indeed, the best-known modern stromatolites are from grazer-free hypersaline lakes in Australia (Awramik 1982). The high salinity of these lakes excludes grazing animals and allows algae to form stromatolites secluded from animals. Borrowing a term from terrestrial paleoecology (Owen-Smith 1987), metazoan grazers can be thought of as "keystone herbivores." Without their efforts, the algal blanket might return to the sea floor and cause a return of the ecological conditions that prevailed during and before the Garden of Ediacara.

DOUBLE FEEDBACK LOOPS

FIGURE 8.11 diagrams the feedback loops that led to the Cambrian explosion. Both extrinsic (environmental) and intrinsic (due to the animals themselves) factors are addressed, and they are inextricably linked in the creation of a more complex marine biosystem, through food chain lengthening and increase in the height of the trophic pyramid. For instance, as available environmental niches are filled, more biotic resources are available and a new predator can evolve to take advantage of the newly evolved prey. These changes are fed back into the evolution of the organisms themselves. Skeletonization and other defenses such as burrowing can occur in response to increases in predation pressure. Burrowing can uncover new sources of food and change the sea floor in ways beneficial for other heterotrophic animals. Behavioral changes in animals can occur via both predator-prey escalation and improved efficiency of low-level heterotrophic feeding strategies, for instance, as when deposit feeders

	Garden of Ediacara	Heterotrophic Pyramids
Dominant trophic strategies	Autotrophy Filter feeding Passive nutrient absorption	Predation Grazing Invasive deposit feeding
Dominant animal body plan	Radial	Bilateral
Typical nervous system	Simple nerve nets ennervating large surface areas	Brains and image-forming eyes

FIGURE 8.10. A comparison of the major features of the Garden of Ediacara-style of ecosystem with those of the familiar Heterotrophic Pyramid ecosystem which has dominated the earth since the Cambrian.

burrow in a geometrically regular pattern to avoid excavating the same sediment over again. When brains reach a certain complexity, innovative behaviors can appear (Calvin 1986).

In a study examining the evolutionary increase in brain size beyond the minimal amount needed to support a given body size, Russell (1981) argued that progressive encephalization appears to have been an evolutionary process with a fairly regular pattern for the last 600 million years. More interestingly, Russell (1981) hypothesizes that the rate of encephalization corresponds to the availability of marine nutrients. He suggested that the establishment of land life during the later Paleozoic may have impaired the flow of terrestrial nutrients to the oceans, producing a lowering (at approximately the Permo-Triassic boundary) of the rates of maximum increase of encephalization on the planet. Russell's suggestion, if confirmed, would indicate a positive correlation between encephalization rate and nutrient supply, analogous to the positive correlation suggested here between the prevalence of hard path strategies and marine nutrient availability. Russell's hypothesis would fit well with the double feedback loops shown in figure 8.11: hard path strategies high on the

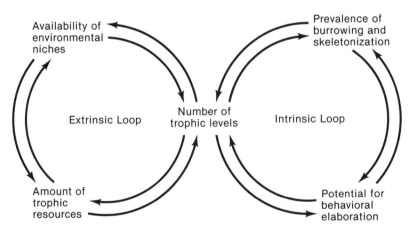

FIGURE 8.11. Double feedback loop showing linkages between four factors that contributed to increases in the number of trophic levels during the Precambrian-Cambrian transition. Extrinsic factors are environmental/ecological conditions, and intrinsic factors are determined by the biology of the organisms which constitute the higher trophic levels of an ecosystem. This figure-eight causal loop diagram may have been initiated by changes in ocean chemistry that increased the amount of trophic resources available in shallow marine waters.

trophic pyramid and behavioral elaboration (intelligence) seem to be linked on the intrinsic loop; both are dependent on the availability of trophic resources as shown on the extrinsic loop.

It must be stressed that all these feedback loops were initiated during a time of a transgression, supercontinental breakup, and unprecedented ocean chemistry changes. Implicit in the double loop diagram of figure 8.11 is the idea that the entire system is ultimately driven by the extrinsic loop. In other words, biological innovations are critically dependent on nonliving supplies of nutrients and other conditions needed for life to flourish. Both the extrinsic and intrinsic loops affect each other, but the extrinsic loop has a much greater effect on the intrinsic loop than vice versa.

The creation of new levels of diversity and new trophic levels was most intense during the Early Cambrian and has been less so ever since, so that, although the "figure-eight" diagram of figure 8.11 involved positive feedback, it is not an example of runaway positive feedback. Clearly, factors after the Early Cambrian must have forced a decrease in the rate of creation of new trophic levels, because a half billion years later, there are still only a limited number of these niches. Here again, we witness the primacy of the extrinsic loop; animals quickly filled relatively "easy" new niche possibilities, and as these became filled, newer niches became successively more difficult to colonize, primarily because of environmental boundaries and the increasingly hostile character of the uninhabited territory yet remaining (i. e., severe wetting and drying conditions on the shoreline, lack of food and oxygen in deep sub-seafloor sediments). Although the process of enrichment and elongation of the trophic heirarchy was slowed after the Cambrian, the process inexorably continued (and continues), eventually leading to the virtually simultaneous colonization of land by animals and vascular plants.

The changes occurring on the extrinsic loop during the late Vendian seem to have favored heterotrophic animals over photoautotrophic animals. We feel that the interactions diagrammed in figure 8.11 first fully came into operation at, or very near, the Precambrian-Cambrian boundary. One change in the physical environment that we have not yet dealt with, but which may have the greatest significance of all, is the role of atmospheric oxygen in the emergence of animals.

IX

Precambrian Oxygen

OXYGEN IS a very unusual gas to find in a planetary atmosphere.
Perhaps the most astute scientific prediction made in the second half of the twentieth century is J. E. Lovelock's surmise that the surface of Mars is currently lifeless because its atmosphere is at chemical equilibrium (Brand, Sagan, and Margulis 1977). The air on Mars is chemically stable and does not show chemical anomalies such as the presence of highly reactive free oxygen. The presence of large amounts of a corrosive gas such as oxygen can only be explained as a result of the activities of gas-producing organisms. Metals will not rust on the surface of Mars, because Mars has an essentially dead atmosphere.

Evidence for accumulation of significant amounts of free oxygen in the earth's atmosphere comes first from sediments deposited well before the Vendian-Cambrian transition. Before 2 billion years ago, free oxygen was rare as an atmospheric gas. Pure iron exposed to the air would not rust. Given enough time, however, it might be dissolved and carried to the sea as native or ferrous iron. We know that oxygen was rare in the pre-2-billion-year atmosphere because minerals such as uraninite (a uranium oxide) and pyrite (iron sulfide) were deposited as common sedimentary particles in clastic sedimentary rocks. These minerals are seldom, if ever, found as sedimentary fragments in later strata, because reaction with oxygen in the atmosphere chemically destroys these minerals long before they have a chance to become incorporated into a sedimentary rock as a constituent grain.

There is consensus among geologists that atmospheric oxygen was limited before 2 billion years or so ago, and that oxygen accumulated in the atmosphere as a waste gas, a byproduct of carbon dioxide/water based photosynthesis (Walker 1987). Oxygen currently constitutes about 21 percent of the earth's atmosphere, but was essentially absent from the atmosphere before some 2 billion years ago. The only other statement about the history of atmospheric oxygen accumulation that can be made with any confidence is that it is highly unlikely that current concentrations of oxygen have fallen much below their present values for some hundreds of millions of years (Margulis and Lovelock 1986). This is so because the cells of all animals require oxygen for cell division.

A popular view among geologists is that oxygen accumulated gradually after 2 billion years ago and that increments in oxygen levels had profound effects on organisms. According to this view, oxygen levels gradually increased until they stabilized at current atmospheric levels. There is little doubt that organisms produced oxygen before 2 billion years ago, but this oxygen was unable to accumulate as a gas because iron dissolved in seawater combined with the oxygen to form rust (iron oxide), a precipitate that sank, chemically inactive, to accumulate on the sea floor. Just as salt has accumulated in the oceans over billions of years, unoxidized (or reduced) iron was abundant in the seas before 2 billion years ago, and was available to "neutralize" the waste oxygen. Thus, dissolved iron performed an important oxygen disposal service; oxygen is a deadly toxin to organisms that do not have special enzymes to limit its reactivity.

Once the reduced iron was removed from sea water (and precipitated on the sea floor as Precambrian iron formations; much of the iron mined for our automobiles is derived from these formations), oxygen began to accumulate in water and air. Life in the seas was either restricted to environments where oxygen remained rare, or was forced to develop enzymes (some bioluminescent!) capable of detoxifying oxygen. Oxygen could also be used by heterotrophic organisms to "burn" the biologic fuel captured in the form of the bodies of their prey.

Cloud (1976, 1983) links increases in oxygen levels to attainment of critical stages in animal evolution. In doing so, he follows the lead of Berkner and Marshall (1965), who suggested that when oxygen in air and water reached a critical level of about 10 percent of the present level of oxygen, the Cambrian radiation occurred—because sufficient oxygen levels then (and only then) allowed animals to radiate. In addition to being a requirement for cell division, oxygen is needed for the synthesis of collagen, an important protein that forms connective tissue in all animals (Towe 1981). According to this scenario, insufficient oxygen availability was the reason that animals appeared when they did, rather than much earlier.

This "oxygen threshold" hypothesis is, however, open to question. Surely Vendian, and—if present—Riphean, animals had both cell division and collagen. Schopf's (1981) "best guess" for the time of full oxygenation of the atmosphere was at 2 billion years ago. But James C. G. Walker (University of Michigan) does not easily dismiss the gradual accumulation model. Walker writes (1987, personal communication):

> I used to believe that the increase in oxygen would have been rapid once the rate of supply of oxygen exceeded the rate of supply of reduced [unoxidized] material and that essentially modern oxygen levels would have been achieved early in the Proterozoic [shortly after 2 billion years ago]. I am now more inclined to favor a gradual increase athough I am not aware of any convincing evidence or even theoretical arguments that illuminate the question. Without such evidence I would be very cautious in attributing advances in Metazoan evolution to increasing atmospheric oxygen.
>
> If the increase was gradual it must almost certainly have reflected a gradually increasing level of global biological productivity. This in turn probably resulted from increasingly efficient organization of the biota. It is hard to see what physical or chemical factor might have limited productivity and gradually have relaxed the limitation. But if we are going to invoke biological improvements to explain the gradual increase in oxygen why not also invoke biological improvements to explain the stages in Metazoan evolution?

Unfortunately, there is no direct and easy way to measure Precambrian levels of oxygen. Aside from the presence or absence of certain minerals such as uraninite and pyrite in sediments, there is as yet no clear way to make a determination of previous oxygen levels using minerals. Discovery of such a technique would be a great boon to studies of ancient life.

It is generally assumed (Pearse et al. 1987; Cloud 1976) that oxygen levels of earth's atmosphere have been approximately the same for the past 100 million years or so. This assumption may have to be abandoned in the face of new research. Marine sediments deposited during the Middle and Late Cretaceous contain unusually large amounts of organic carbon. Some of these deposits are referred to as Cretaceous Black Shales. The reason, or reasons, for the formation of these shales are not certain, but may have to do with the lack of deep water circulation in Cretaceous seas. Most organic carbon is ultimately derived from the activities of photosynthetic marine organisms. The carbon-rich Cretaceous sediments must have been balanced by a surplus of the important byproduct of most photosynthesis, free oxygen. We have calculated that the unoxidized carbon in Cretaceous Black Shales alone could have raised atmospheric oxygen from 21 to 24 percent (McMenamin and McMenamin 1987).

A few months after our estimates were published, direct measurements of the oxygen levels of Cretaceous air were announced by R. Berner of Yale University and G. Landis of the U.S. Geological Survey in Denver. Berner and Landis (1987, 1988) used air bubbles trapped in Cretaceous amber to make their measurements (Kerr 1987). It is surprising that this new and innovative technique was not attempted earlier, because the idea of using bubbles in amber to sample ancient air was first suggested over twenty years ago by a science fiction television program. Frank Grober (1987, written communication) remembers a science fiction program shown back in the late 1950s and early 1960s that dealt with analyzing ancient atmospheres by looking at gas bubbles in amber. (This is not the only instance known of scientific practice catching up with ideas generated in the science fiction literature.)

Berner and Landis' analysis of the trapped gas by quadropole mass spectrometry revealed that the 80 million-year-old amber had trapped air containing roughly 30 percent oxygen versus the 21 percent oxygen of today's air (Kerr 1987). James C. G. Walker has a model of evolving atmospheric oxygen that indicates a 27 percent level of oxygen in the Cretaceous (Kerr 1987). These are startling results, particularly considering that calculations by Watson et al. (1978) indicate that even wet vegetation is dangerously flammable at atmospheric oxygen levels above 25%. Berner and Landis' results have been challenged by those who feel that gases diffuse too quickly to retain the original gas in amber bubbles for millions of years, and others who believe that their technique may be invalid (Hopfenberg et al. 1988). Regardless of what the actual values of Cretaceous oxygen were, it is important that both sedimentological research and other studies indicate that oxygen levels have been higher in the geologic past than they are at present, a possibility that was first considered by J. C. G. Walker in 1974.

Much research has focused on lowered levels of atmospheric oxygen during the Precambrian. The other alternative, that oxygen levels were *higher* at times during the Precambrian than at present has not been much discussed. Once the "sinks" for free oxygen, such as dissolved iron, were saturated, there is little that would have prevented oxygen levels in the Precambrian from getting much higher than they are today. This is particularly so since there is no evidence for the presence of Precambrian land plants which could have acted as a negative feedback for continued increases in oxygen levels: Forest fires combine oxygen with organic hydrocarbons to form carbon dioxide, and as Watson et al. (1978) showed, the likelihood of these fires increases with increasing levels of oxygen.

Study of carbon isotopes may give indirect clues to the levels of oxygen in the Riphean and Vendian. As discussed in the last chapter with regard to sulfur isotopes, organisms tend to use the most energy-efficient metabolic processes, and that means preferentially utilizing the lighter of two isotopes of the same element. Two stable, naturally occurring, isotopes of carbon are ^{12}C (light carbon) and ^{13}C (heavy carbon). (Radioactive carbon or ^{14}C is another isotope that is extremely useful for dating fossils less than about 70,000 years old.)

According to Knoll et al. (1986), Riphean calcium carbonate sediments of Greenland and Svalbard have elevated levels of heavy carbon. Since inorganic sedimentary rocks are composed of the carbonate ions "cast off" by organisms, a bias toward heavy carbon in the inorganic minerals (which precipitate from nonliving seawater) can indicate widespread organic productivity and widespread removal of light carbon from sea water and incorporation of this light carbon into organic hydrocarbons. Since photosynthetic production of hydrocarbons is accompanied by release of oxygen, Knoll et al. (1986) suggest that the elevated ^{13}C in Riphean limestones and dolomites

165

may indicate higher values of oxygen in the Riphean. The data are open to several interpretations, however, and it is not possible to tell how high the oxygen levels were at that time. But they may very well have been as high or higher than present levels. For this reason, we think that the oxygen threshold hypothesis is probably wrong.

There is good reason to expect high levels of organic accumulation in sediments during the Riphean. Bioturbation was limited and there were abundant benthic moneran communities, both of which would lead to production and preservation of buried organic matter and to increased levels of atmospheric oxygen (Fischer 1984; Lambert et al. 1987). Lambert et al. (1987) suggest that these sequences contain enough organic material to have produced important late Precambrian oil and gas deposits.

Another factor of potentially great significance for Riphean-Vendian oxygen and the global carbon cycle are methane hydrates (sometimes called gas hydrates). Methane hydrates are bizarre ice-like compounds that form in sea floor sediments and under permafrost whenever methane gas and water are present and temperature and pressure conditions are appropriate (Kvenvolden and McMenamin 1980). The amount of carbon (in the methane molecules) trapped in hydrates today is enormous. The hydrate reservoir may trap an amount of carbon that rivals the amount present in all oil and gas deposits throughout the world (about 10,000 gigatons; Kvenvolden 1987). During the severe Varangian glaciation, the temperature-pressure zone favorable for hydrate formation may have expanded in land and sea sediments, trapping gigantic quantities of gaseous carbon in addition to the carbon trapped in sediment as solid organic compounds. The colder it became, the broader was the field of methane hydrate stability, the more methane became trapped and prevented from reaching the atmosphere, and the more oxygen was free to accumulate in the atmosphere.

The case that we make here for higher oxygen levels in the Precambrian is speculative. We offer it primarily to provide an alternative to the idea that low oxygen level was the major factor in controlling the timing of the Cambrian animal radiation, a hypothesis we find unconvincing.

X

Ecological Chaos and Innovation

Evolution is chaos with feedback.
JOSEPH FORD

THROUGHOUT THIS book we have referred to the Cambrian radiation of animals as an "explosion," a term which implies that the appearance of these animals was a geologically sudden event. Not all paleontologists agree with this, and some prefer to see the Cambrian radiation as the moment in geologic time when animals first appear as shelly *fossils*, not necessarily when these types of animals first came into being. Therefore, there are two ways of viewing the Cambrian explosion: either it was the nearly simultaneous evolution of a number of animal phyla (the "bang" hypothesis); or, Cambrian animal phyla had long Precambrian histories that were not recorded as fossils (the "whimper" hypothesis). In other words, did the Cambrian come in with a whimper or a bang?

The whimper hypothesis has one major point in its favor. There were animal trace fossils occurring well before the Cambrian boundary, and arthropods (Jenkins 1988) and perhaps even echinoderms (Gehling 1987) are known from Vendian strata. One line of argument, however, acts as a fatal blow to the whimper hypothesis. With no more than five known exceptions, all of the extant well-skeletonized animal phyla first appear as fossils in the Cambrian. Also, many phyla that are now extinct first appeared in the Cambrian. As many as 100 phyla may have existed during Cambrian, and only 5 percent or less of this number show any evidence of a Precambrian ancestry. Thus, the genesis of most animal phyla must date back to the Cambrian and not before then.

The arguments above are largely based on the work of the paleontologist James W. Valentine. In the late 1960s, Valentine realized that large numbers of phyla burst upon the scene during the Cambrian evolutionary radiation, and that virtually no new phyla appeared during the many radiations of animals that followed the Cambrian (Valentine 1969). This effect also held true for other higher taxa of animals.

The biological taxonomic hierarchy consists of seven basic levels: kingdom, phylum, class, order, family, genus, species. The kingdom level contains the five-fold division of all life: monerans, protists, fungi, animals, plants. Along with kingdom, the phylum, class, and order levels are considered to be higher taxa; the family, genus, and species levels are considered to be lower taxa. Valentine (1969) noted that not only did the Cambrian contain most of the originations of animal phyla, but also it had a disproportionate share of class-level originations. For example, snails, clams, cephalopods, monoplacophorans, and rostroconchs, representing five different classes within the mollusk phylum, *all* first appear in the Early Cambrian.

Why did all these higher taxa emerge during the Cambrian radiation? Two principal hypotheses have been offered over the years; the conventional ecological hypothesis and the genomic hypothesis.

According to the conventional ecological hypothesis, major evolutionary innovations could occur during the Cambrian because marine diversity was low and there existed a limited level of competition (Gillett 1985). Such an ecological setting offered greater opportunity for the establishment of evolutionary innovations that might not have survived in an ocean with more of its ecological niches filled.

The genomic hypothesis holds that the genetic programs of animals were more easily changed early in metazoan history (Valentine and Erwin 1987). As time passed, it became increasingly difficult to make significant changes in the animal genetic codes: they became "locked in" on particularly successful sequences of development and less forgiving of major restructuring. This "locking in" effect is often called canalization, by analogy with a stream cutting its channel so deeply that it becomes increasingly difficult to switch the stream flow to another direction. Since the genomic hypothesis implies that major mutational change in animals could only occur when genes were young and unsophisticated, we like to call it the "green genes" hypothesis.

Both the conventional ecologic and the green genes hypotheses attempt to explain why the Cambrian explosion didn't (or couldn't) happen again. After the devastating Permo-Triassic mass extinction about 200 million years ago, as many as 96 percent of all marine animal species went extinct. Erwin et al. (1987) calculate that after this mass extinction, animal lower taxa were reduced to Vendian levels of diversity. In spite of this, there was no repeat of the Cambrian explosion. No new phyla appeared during in the Mesozoic. Under the green genes hypothesis, no new Mesozoic phyla appeared because animal genetic codes were not as flexible nor as able to undergo radical change as they were in the Cambrian. With the conventional ecologic hypothesis, which Erwin et al. (1987) apparently lean towards, no new phyla appeared after the Permo-Triassic mass extinction because nearly all of the Paleozoic phyla had survived. The Mesozoic marine world was already "seeded," in Noah's Ark fashion, with representatives of all the basic types of animals, which quickly radiated to repopulate the extinction-ravaged marine world of the early Mesozoic.

The green genes and conventional ecological hypotheses both have serious drawbacks that, in our opinion, show that neither can adequately explain the Cambrian explosion. The main problem with the conventional ecological hypothesis concerns early land animals. Animal life on land and in fresh water dates back to the Ordovician (Retallack and Feakes 1987), and possible nonmarine arthropods such as *Kodymirus* (figure 10.1) date back to the Early Cambrian (Gray 1988). Before this time, land environments were presumably empty of animals, and offered a very open and uncrowded ecological setting. No new phyla, however, appeared on land, during a time early in animal history when animal genes might be expected to retain much of their Early Paleozoic "green genes" flexibility. In fact, relatively few animal phyla were ever able to become very successful on land or in freshwater environments. This failure to reproduce the taxonomic results of the Cambrian explosion, under wide-open ecological conditions and without previous "seeding" of the land environment with animal taxa, is the downfall of the conventional ecological hypothesis.

It is true that the land environment is more physically challenging than the marine environment. Temperature fluctuations are greater, and life for animals and plants requires structural support and tricks to prevent dehydration. But these difficulties are offset by the fact that the land offers more in the way of primary food production. Land plants, even the earliest ones, had access to more energy than their aquatic ancestors because sunlight is more intense in air than under water. Further, the animals that did solve the problems of life at the bottom of the air (as opposed to bottom of the sea) were able to diversify with great success. There is no reason why the physical rigors of a terrestrial environment alone would have prevented the evolution of new phyla on land. Under the conventional ecological hypothesis, one might have *expected* the appearance of completely new types of animals to cope with land's physiological challenges.

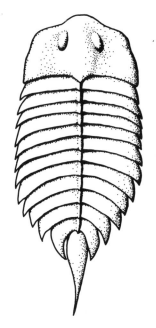

FIGURE 10.1. *Kodymirus vagans,* a·Lower Cambrian arthropod from Bohemia that may have lived in non-marine environments. Length of specimen 12.7 cm. (After Chlupáč and Havlíček 1965)

On the other hand, the green genes hypothesis fails on the fact that animal trace fossils are known from strata deposited tens of millions of years (and perhaps as many as 50 to 150 million years) before the Cambrian boundary. The genomic hypothesis is at a loss to explain why Vendian metazoans, surely with very green genes, failed to undergo a major radiation millions of years before they finally did.

The viewpoint presented in this book, which could be called the Garden of Ediacara hypothesis or the unconventional ecological hypothesis, avoids the problems encountered by the conventional ecological and green genes hypotheses. The new hypothesis shares some aspects of the conventional ecological hypothesis. For instance, an uncrowded Cambrian environment plays a role in the Garden of Ediacara hypothesis. The crucial difference between the conventional and the unconventional ecological hypotheses lies in how each hypothesis characterizes the duration of the underpopulated interval in ocean history. In the conventional hypothesis, marine environments waited patiently and passively for three billion years before animals finally arrived to fill up the blank habitats or niches. This scenario might be plausible if the Cambrian explosion was close to the origin of multicellular organisms (i.e., within a few million years), but we know that it was not.

In the Garden of Ediacara hypothesis, the oceans were never truly ecologically empty; the Precambrian marine world was *not* vacant and passively waiting for animals to appear, but rather was a vital world filled to the brim with soft path organisms. This ecosystem was destroyed and displaced by the origin of a multitrophic level, hard path system dominated by heterotrophic animals. Animals present before the Cambrian seem to have lived by Precambrian ecological "rules"; hence the body plans of Ediacaran animals that were shaped for soft path life styles.

The Garden of Ediacara was overthrown at the Cambrian boundary by heterotrophic communities in a rather short period of time. Using the speedy appearance of abundant shelly fossils as a gauge (Conway Morris 1988), we estimate that the turnover took only a few million years, rapidly and radically changing the basic ecology of the marine biosphere. Immediately after the "fall" from the Garden of Ediacara, animal diversity was still relatively low but began to build up rapidly, immediately thereafter—this is what we call the Cambrian explosion. It is important to note that low taxonomic diversity does not necessarily imply low biotic competition, as required by the conventional ecological hypothesis.

An important feature of the Garden of Ediacara hypothesis is that it can explain the taxonomic lopsidedness of the Cambrian explosion, or why so many higher animal taxa appeared at that time in comparison to radiations before and after. The essential question is this: How do you evolve from from, say, a worm to a brachiopod or flatworm to a clam in a geological second—a mere one million years? Erwin et al. (1987) note that there is not nearly enough geological time available for Cambrian phyla to appear by the gradual, slow accumulation of relatively minor evolutionary changes. Whatever caused the Cambrian animal phyla to appear caused them to appear quickly.

We suggest that the predators that came as part of the Cambrian hard path revolution played a crucial role in creation of Cambrian higher taxa such as new phyla. Many of the animals that were able to defend themselves against predators (by burrowing deeply, forming shells, and other strategies) *incidentally* found themselves in a position to capitalize on food resources that had been weakly exploited or not exploited at all by animals during the Precambrian. Shells, for instance, greatly aided both filter feeders (e.g., brachiopods, eocrinoids) and deposit feeders (e.g., trilobites, who used their arched carapaces as a sediment-processing factory; Seilacher 1985). Deep burrowers excavated food resources unavailable before. In chapter 8, we examined the explosive potential for innovation and feedback inherent in larger brains. Cambrian predators forced other animals to make the jump to successful lifestyles which, if no one had been chasing them, they might never have attempted. Probably the only jumps that were successful were those that were made quickly, while ecological conditions on the sea floor were in a state of turmoil.

The Garden of Ediacara hypothesis also explains why no new animal phyla appeared with the animal invasion of land. Land colonization by animals occurred some time between 50 to 100 million years after the Cambrian radiation, and the successful first colonizers established a multi-trophic level hard path ecosystem inherited from the Cambrian radiation. Early land animal fossil communities are rich with arthropod predators, so rich that in one well-preserved early terrestrial fauna, all but one of the identifiable components of the fauna are predators (Shear et al. 1984). Opportunities for major evolutionary innovation and the evolution of new phyla were present on land before colonization, but these opportunities were decisively cut off when the Cambrian ecosystem was transposed onto land and freshwater environments. Most importantly, this migration of an ecosystem was accomplished with only a very few phyla (initially perhaps only two: arthropods and annelid worms). According to the tenets of the conventional ecological hypothesis, these two phyla on land should not have prevented evolutionary innovation there. After all, worms—and probably arthropods—were around during the Vendian, but they did not hinder the Cambrian explosion.

One might validly argue, on ecologic grounds, that if land colonization by animals had occurred earlier, say at the beginning of the Cambrian explosion when animal phyla were originating and before ecological hierarchies became fixed, there might have been a better chance of evolving distinctive new body plans and radically new types of living on land. Land animals are, after all, relatively minor modifications of body plans that evolved in the sea during the Cambrian explosion. The bodies of the land fauna remain well-suited for marine life; indeed, many terrestrial animal groups (whales and other cetaceans, sea turtles, ichthyosaurs, penguins, sea snakes, some insects) have returned to the sea over geologic time. The best way to explain the lack of new phyla on land is that early land faunas were initiated via a hard path ecosystem, which quickly choked off any possibility of a truly new fauna evolving on land. This would have been true regardless of which animal phyla had first colonized land during the Ordovician, since they would have been products of the heterotrophic ecosystem.

The Cambrian explosion is a geologically abrupt event of great significance. It represents the collapse of an old marine ecosystem dominated by autotrophs and filter feeders and its replacement by a more complex ecology dominated by multilevel trophic pyramids. The apparent suddenness of the event (at most a few million years) is real, and new feeding types and anatomical designs of animals rapidly appeared. The old order vanished, or was driven to marginal environmental situations, and a new order was quickly established. Many of the groups which are most successful in the post-Cambrian oceans achieved their dominance as a consequence of good fortune during a time of profound ecological change. Skeletonization proved to be a successful strategy for coping with these ecological changes, but was by no means the only way. Vendian anatomies and behaviors that were able to capitalize on, or fruitfully coexist with, emerging hard path feeding styles and deal with changes in the seafloor substrate rapidly gained an upper hand. The characteristics of oceanic ecosystems haven't changed much since, and the younger land ecosystems are also much the same—in fact, more similar than one might expect if one considers the environmental differences between land and sea.

The arguments above imply that most animal phyla came into being during a brief episode of ecological chaos during the transition between the Garden of Ediacara and the subsequent ecological system. There appear to have been opportunities opened up during this time of ecological turmoil which have remained unavailable during less chaotic times. Evolution as "chaos with feedback" (Ford, quoted in Gleick 1987:314) must have been particularly true during the Vendian-Cambrian boundary; in fact, the feedback processes sketched in figure 8.11 may operate fully *only* during times of major ecological chaos. The barriers to innovation, that exist in the structural checks and balances of an ordered ecosystem, are temporarily opened during the formative stages of a new ecosystem. After the new system is set, the doors to major innovation may be closed until the next episode of ecosystem disruption, although minor innovations and adjustments can occur at any time. Feedback such as predator-prey escalation occurred after the Cambrian explosion (Vermeij 1987), but was comparatively weak and ineffectual at bringing about major new innovations. The greatest innovations occur during the most severe episodes of ecologic change.

The results of the Cambrian ecological revolution are now an integral part of life. All features of the living world, from leaf-mimicking insects to human intelligence, bear the mark of our "fall" from the Garden of Ediacara. The current ecological system has been ascendant for half a billion years. This is no guarantee, however, that the current heterotrophic ecosystem will always be the dominant style of ecological interaction on earth. Perhaps we human beings and our inventions are already playing a role in the next ecological metamorphosis.

To conclude, the division between the Cambrian and the Precambrian marks an event of primary importance to the history of life, and this boundary should be recognized as the major division in the geological time scale. The Vendian-Cambrian transition was a strange and unsettled time. The ecological changes of this interval determined much of subsequent earth history. Environmental and ecological constraints on the evolution of animals were minimal during the transition, and the feedback-driven flood of innovation that followed resulted in the appearance of most familiar animal phyla. The

Precambrian-Cambrian transition was the unique conjunction of burgeoning heterotrophs and a collapsing Garden of Ediacara milieu. This unprecedented ecological transformation permitted a breathtakingly rapid rate of animal evolution which could not be supported during later, less chaotic times.

APPENDIX

Selected Specimens

This appendix lists locality and stratigraphic positions plus repository information for specimens which are illustrated in this book but which have not been published elsewhere; see also figure 4.30 and M. McMenamin (1984) for information on the specimens from Sonora, Mexico. Specimens deposited in the Institute of Geology Museum (address: Departmento de Paleontología, Instituto de Geología, Cuidad Universitaria, Delegacíon de Coyoacán, 04510, México, D. F.) are assigned four-digit Institute of Geology Museum (IGM) numbers. GSC refers to the Geological Survey of Canada. (Dr. Thomas E. Bolton, Geological Survey of Canada, 601 Booth Street, Ottawa KIA OES, Canada). UCMP refers to the University of California Department of Paleontology, Berkeley, California 94075.

FIGURE 3.4. Pit-and-mound structure, from within the top 14 meters of the Clemente Formation, Cerros de la Ciénega, Sonora, Mexico. Sample MM-82–79b; IGM 3630.

FIGURE 3.5. *Skolithos linearis*, Proveedora Quartzite (float from basal part of unit), Cerro Clemente, Sonora, Mexico. Sample 3 of 12/15/82; IGM 3633.

FIGURE 3.6. *Scolicia* sp., uncollected float block from the middle part of unit 3, Puerto Blanco Formation, Cerro Rajón Area, Sonora, Mexico.

FIGURE 3.7. *Cruziana semiplicata*, from within the lower 15 meters of the middle shaly interval of unit 3 of the Puerto Blanco Formation, Cerro Rajón area, Sonora, Mexico. Sample 7 of 12/17/82; IGM 3622.

FIGURE 3.9. Star-shaped trace fossil, uncollected float block from the middle part of unit 3, Puerto Blanco Formation, Cerro Rajón Area, Sonora, Mexico.

FIGURE 3.10. *Protopaleodictyon* sp., from quartzite beds within the Gog Group, Lake Louise, Alberta, Canada. (Collected and photographed by J. P. A. Magwood.)

FIGURE 4.5. *Hyolithellus* sp., from the lower archaeocyathan limestone part of unit 3 of the Puerto Blanco Formation, Cerro Rajón area, Sonora, Mexico. Sample MM-82–49; IGM 3614(8).

FIGURE 4.8. *Lapworthella filigrana*, mitrate sclerite, from a 1 cm thick limestone bed in the lower part of the middle shaly part of unit 3 of the Puerto Blanco Formation, Cerro Rajón area, Sonora, Mexico. Sample 7c of 12/17/82; IGM 3642.

FIGURE 4.19A. *Lingulella* sp., from the lower archaeocyathan limestone part of unit 3 of the Puerto Blanco Formation, Cerro Rajón area, Sonora, Mexico. Sample MM-82–49.

FIGURE 4.19B. *Mickwitzia* sp., from 59 meters above the base of unit 2 of the Puerto Blanco Formation, Cerro Rajón area, Sonora, Mexico. Sample 5.5+ of 12/17/82; IGM 3614(33).

FIGURE 4.19C. *Kutorgina* sp., from 26 meters above the base of unit 2 of the Puerto Blanco Formation, Cerro Rajón area, Sonora, Mexico. Sample 4 of 12/17/82.

FIGURE 4.19D. *Paterina* sp., from 26 meters above the base of unit 2 of the Puerto Blanco Formation, Cerro Rajón area, Sonora, Mexico. Sample 4 of 12/17/82.

FIGURE 4.21. *Pelagiella* sp., Buelna Formation, Angustura Pass area, Sierra del Viejo, Sonora, Mexico. Sample MM-82–83.

FIGURE 4.22. *Bemella* sp. Cassiar Mountains, Northwest Territories, Canada. Sample 96897. GSC Specimen Number 95384.

FIGURE 4.28. *Microdictyon* sp., from unit 2 of the Puerto Blanco Formation, Cerro Rajón area, Sonora, Mexico. Sample MM-82–49.

FIGURE 4.29. Ostracode, from the lower archaeocyathan limestone part of unit 3 of the Puerto Blanco Formation, Cerro Rajón area, Sonora, Mexico. Sample MM-82–49.

FIGURE 7.7. Bored *Bemella* sp., from 59 meters above the base of unit 2 of the Puerto Blanco Formation, Cerro Rajón area, Sonora, Mexico. Sample 5.5 of 12/17/82.

FIGURE 7.8. Bored *Hyolithellus* sp., from the lower archaeocyathan limestone part of unit 3 of the Puerto Blanco Formation, Cerro Rajón area, Sonora, Mexico. Sample MM-82–49.

FIGURE 7.9. *Mickwitzia* sp., shales in lower part of middle of the Poleta Formation, Blanco Mountain quadrangle. UCMP Locality Number B-9852; on southeast slope of hill 7821 in southern part of Sec. 5 and northern part of Sec. 8; collected from outcrops extending from middle of NW1/4 NE1/4 NW1/4 Sec. 8 to middle of SE1/4 SE1/4 SW1/4 Sec. 5, T. 8 S., R. 35 E. Collected by J. W. Durham and R. Gangloff in 1962; UCMP specimen number pending.

FIGURE 8.6. Casts of endolithic borings on a partially dissolved echinoderm plate, Buelna Formation, Angustura Pass area, Sierra del Viejo, Sonora, Mexico; Sample MM-82–83.

FIGURE 8.5. Flat pebble conglomerate (bedrock exposure, not collected), unit 1 of the Puerto Blanco Formation (102 meters below the contact between units 1 and 2 of the Puerto Blanco Formation), Cerro Rajón area, Sonora, Mexico.

GLOSSARY

acritarch Any small, roughly spherical fossil with a wall composed of an organic material such as sporopollenin; usually removed from enclosing sediments by maceration (dissolving the rocks with hydrofluoric acid).

Animalia The kingdom to which animals belong.

anoxic Depleted in oxygen, as are deep marine waters with poor circulation.

aragonite A mineral, composed of calcium carbonate, which forms needle-shaped or fibrous crystals.

arthropod A member of the phylum Arthropoda, animals with external skeletons and jointed appendages; the phylum includes crabs and insects.

autotrophic Capable of biochemically manufacturing food from simple inorganic chemicals (c.f. *heterotrophic*).

basalt A dark-colored, fine-grained igneous volcanic rock. Basalt is the major igneous rock erupted during rifting and forms a major portion of the oceanic crust.

benthic   Living on the sea floor or lake bottom.

bioclastic   Formed from fragments of broken shell. C.f. *clastic.*

biomass   The total mass or weight of all organisms in a particular region or category.

biota   All the life forms in a region or a given interval of geological time.

Burgess Shale   The formation bearing a famous find of Middle Cambrian soft-bodied and shelly fossils; the fauna was discovered by C. D. Walcott in British Columbia, Canada.

calcite   A mineral composed of calcium carbonate which forms prismatic crystals.

Cambrian   The first period of the Paleozoic; see figure 1.1.

cephalon   The head region of a trilobite, where its eyes and glabella are located; see figure 4.12.

cephalopod   A member of the mollusk class Cephalopoda, which includes squid, octopi, and ammonites.

cerata   Horn- or finger-like projections on the outer surface of a soft-bodied animal which increase the animal's total surface area or exposure to light.

chemoautotrophic   Capable of using simple molecules such as hydrogen sulfide or methane to biochemically manufacture food. C.f. *photoautotrophic*

chemosymbiosis   A biological relationship in which a chemoautotrophic organism (the chemosymbiont) lives within the tissues of larger host organism. The host receives nutrients and waste removal services from the chemosymbiont and the chemosymbiont receives protection from the tissues of the host.

chetognath   A phylum of small marine predatory metazoans; also called arrow worms.

chloroplast   A chlorophyll-bearing organelle able to photosynthesize.

class   A category in the taxonomic hierarchy intermediate between phylum and order.

clastic   Composed of fragments of rocks and mineral grains.

coelom   A rigid, fluid-filled body cavity present in many metazoans; often serves as a hydrostatic skeleton.

coelomate   An animal with a coelom.

community   A group of organisms living together as part of the same trophic pyramid.

crinoid A group of stalked, filter-feeding echinoderms; also called sea lillies.

diversity The number of different *kinds* of organisms. Diversity can be studied for a given fauna or during a given interval of geological time.

echinoderm A phylum of animals, generally with spiny external skeletons; includes starfish, sea urchins, and sea cucumbers.

edrioasteroid Any member of an extinct class of sessile, biscuit-shaped echinoderms.

eukaryote Any organism with a membrane-bound nucleus and organelles. Cf. *prokaryote.*

eutrophic Enriched in nutrients. The term applies to bodies of water.

evaporites Sediments (often including various salts) formed by the evaporation of a body of marine water.

extant Living today; not extinct.

family A category in the taxonomic hierarchy intermediate between class and genus.

fauna The entire animal population (modern or ancient) of a given area or geological time span.

foraminifera Marine protists bearing a shell with one or more chambers, usually composed of calcite.

Fungi The kingdom which includes molds, yeast, mushrooms, and other non-photosynthetic organisms that feed upon organic material and reproduce by means of spores.

genal spine The pointed, posterior corners of a trilobite cephalon.

genera Plural of *genus.*

genus A category in the taxonomic hierarchy intermediate between family and species. The first part (always capitalized) of a species' Latin name is its genus.

glabella The large central lobe on a trilobite's cephalon, which housed the animal's stomach.

Gondwana A Precambrian to Mesozoic supercontinent composed of most of the present-day southern continents; see figure 6.2.

gradualism The notion that major change occurs as the incremental summation of numerous small changes over time.

helicoplacoid An extinct class of echinoderms with a spindle-shaped shell composed of spiralling rows of plates; see figure 4.18.

heterotrophic Dependent on other organisms as food sources.

hydrogen sulfide H_2S, the rotten-egg gas.

hydrostatic Supported by internal water pressure.

infaunal Living within the sediment.

isotope Atoms of the same element with differing numbers of neutrons in the atomic nucleus. Radioactive isotopes are unstable and spontaneously lose protons or neutrons from their nucleus via radioactive decay, and are thus changed into a different isotope (or a different element, if they lose protons).

kingdom The highest category in the hierarchical classification of organisms.

Lipalian interval A hypothetical, worldwide unconformity between Precambrian and Cambrian strata. The concept of this interval is no longer accepted by stratigraphers.

lithification The process by which loose sediment particles are bound into a sedimentary rock.

metazoan Any multicellular animal with organ systems and with cells arranged in two layers in its embryonic stages.

mica Any of a group of siliceous minerals that form flat chip-, plate-, or sheet-forming crystals; examples include the minerals biotite and muscovite.

Mirovia The name, proposed in this volume, for the late Precambrian superocean that surrounded Rodinia.

micron One one-thousandth of a millimeter.

Moneran Any unicellular organism without a membrane-bound nucleus and without organelles; examples include blue-green "algae" and bacteria.

niche The role of an individual species in its community and its environment, including its position in the trophic pyramid, its feeding strategy, habitat, reproductive behavior, etc.

nucleus In biology, the membrane-bound organelle where the most of the genetic material in a cell is located; in physics, the massive part of an atom, composed of protons and neutrons.

oligotrophic Depleted in nutrients; applied to bodies of water.

ontogenetic Occurring during the life and development of an individual organism.

order A category in the taxonomic hierarchy intermediate between class and family.

Ordovician   The period following the Cambrian; see figure 1.1.

organelle   A small membrane-bound intracellular body that performs a specific function for the cell such as respiration or photosynthesis.

Pangaea   A supercontinent composed of all or nearly all the continents; it existed during the late Paleozoic and the early Mesozoic.

period   A major unit of geological time; see figure 5.1.

peristaltic burrowing   A type of burrowing which involves rhythmic expansion and contraction of a coelom or other hydrostatic skeleton.

photoautotrophic   Capable of creating food from simple chemicals such as water and carbon dioxide, using photosynthesis.

photosymbiosis   A biological relationship in which a photoautotrophic organism (the photosymbiont) lives within the tissues of larger host organism. The host receives nutrients and waste removal services from the photosymbiont and photosymbiont receives protection from the tissues of the host.

phyla   Plural of *phylum.*

phylum   A category in the taxonomic hierarchy intermediate between kingdom and class.

Plantae   The plant kingdom. Includes all photosynthetic organisms that are neither monerans or protists.

pneumatic   Filled with compressed air or water.

Precambrian   An informal term for the five-sixths of earth's history prior to the Cambrian.

Proterozoic   A late Precambrian interval of geological time from 2 billion years ago to the end of the Precambrian, about 600 million years ago; see figure 1.1.

prokaryote   Any unicellular organism without a membrane-bound nucleus and without organelles; a moneran.

Protista   The kingdom containing unicellular and colonial eukaryotes.

radiation, adaptive   Expansion in numbers or different types of any group of organisms; rapid diversification.

range   The time or stratigraphic interval marked by the first and last appearance of a particular fossil taxon.

rifting   The geological sundering of a continent or supercontinent into one or more smaller continental fragments.

Riphean   An interval of Precambrian geological time before the Vendian; see figure 1.1.

Rodinia   The name, proposed in this volume, for the late Precambrian supercontinent composed of nearly all the continents; see figure 6.1.

scleritome   An external skeleton composed of numerous small hard parts or sclerites.

sessile   Spending part of its life cycle in one place; stationary.

silica   Silicon dioxide, the chemical constituent of the mineral quartz.

species   A group of organisms that can interbreed (or, from fossil evidence, are inferred to have been able to interbreed); the lowest and fundamental unit of classification in the taxonomic hierarchy, ranking next below genus.

sporopollenin   An acid-resistant organic compound (a carotenoid ester) secreted by some cyst-forming organisms.

stage   A subordinate geological time-rock unit; see figure 5.1.

stratotype   A formally defined stratigraphic boundary identified as a point in a specific sequence of strata.

stromatolite   A mound-shaped, cone-shaped or branching, concentrically-laminated structure that is, or was once, an aquatic community of monerans and/or protists; illustrated in figures 1.3 and 1.4.

subtidal   Occuring on the sea floor in water deep enough to avoid exposure during low tide.

supercontinent   Any ancient continent formed of two or more present-day continents.

superocean   Any ancient ocean larger than the modern Pacific Ocean.

system   A major geological time-rock unit. See figure 5.1.

taxa   Plural of *taxon.*

taxon   A given group of any taxonomic rank, such as a particular species or a particular phylum. Kingdom, phylum, and class are higher taxa; family, genus, and species are lower taxa.

taxonomy   The system by which organisms are named and grouped in a hierarchy according to their biological relationships; also called systematics.

terrestrial   Living on land or in freshwater environments.

thin section   A slice of rock mounted on a glass slide and ground thin enough (usually about 30 microns thick) to transmit light and be viewed through a light microscope.

transgression   Rise in the level of the sea with respect to the level of the land. The shoreline will necessarily move inland during a transgression.

trilobite   A member of an extinct class of arthropods which were most abundant during the Cambrian. Illustrated and diagrammed in figure 4.12.

trophic pyramid   A handy way to visualize ecological relationships. Organisms at each level (except the base) eat organisms at the level below. Autotrophs are at the base, herbivores and herbivore predators in the middle, and a heterotrophic top carnivore, or keystone predator, at the top. The relative width of each level represents the total biomass of the organisms at that level in the trophic hierarchy.

unconformity   A major gap or hiatus in a stratigraphic sequence.

vagile   Capable of moving around; mobile.

Vendian   A geological period at the end of the Precambrian; see figure 1.1.

Vendozoa   Seilacher's (1984, 1985) proposed kingdom consisting of the soft-bodied fossils of the Ediacaran fauna.

vent fauna   A community of animals living in deep marine environments where dissolved gases emitted by hydrothermal vents are the primary energy and food source for animals with chemosymbionts.

zone   A subordinate unit in biostratigraphic subdivision of strata.

REFERENCES

Alpert, S. P. 1977. Trace fossils and the basal Cambrian boundary. *Geological Journal Special Issue* 9:1–8.

Alpert, S. P. and J. N. Moore. 1975. Lower Cambrian trace fossil evidence for predation on trilobites. *Lethaia* 8:223–230.

Anderson, M. and S. Conway Morris. 1982. A review, with descriptions of four unusual forms, of the soft-bodied fauna of the Conception and St. John's Groups (late Precambrian), Avalon Peninsula, Newfoundland. *Third North American Paleontological Convention, Proceedings* 1:1–8.

Anderson, R. N. 1985. Sulphur-eating tube worms take to Oregon breaches. *Nature* 314:18.

Aronson, R. B. and H.-D. Sues. 1987. The paleoecological significance of an anachronistic ophiuroid community. In W. C. Kerfoot and A. Sih, eds., *Predation—Direct and Indirect Impacts on Aquatic Communities*, pp. 355–366. Hanover, N.H.: University Press of New England.

Awramik, S. M. 1982. The origins and early evolution of life. In D. G. Smith, ed., *The Cambridge Encyclopedia of Earth Sciences*, pp. 349–362. Cambridge: Cambridge University Press.

Awramik, S. M., D. S. McMenamin, Yin Chongyu, Zhao Ziqiang, Ding

REFERENCES

Qixiu, and Zhang Shusen. 1985. Prokaryotic and eukaryotic microfossils from a Proterozoic/Phanerozoic transition in China. *Nature* 315:655–658.

Awramik, S. M. and R. Riding. 1988. Role of algal eukaryotes in subtidal columnar stromatolite formation. *Proceedings of the National Academy of Sciences (USA)* 85:1327–1329.

Balsam, W. C. and S. Volgel. 1973. Water movement in Archaeocyathids: Evidence and implications of passive flow in models. *Journal of Paleontology* 47:979–984.

Barrande, J. 1852–1911. *Système Silurien du centre de la Bohême* (8 volumes, 29 parts). Prague: Bellman.

Bassler, R. S. 1941. A supposed jellyfish from the pre-Cambrian of the Grand Canyon. *Bulletin of the United States National Museum* 89:519–522.

Bengtson, S. 1970. The Lower Cambrian fossil *Tommotia*. *Lethaia* 3:363–392.

Bengtson, S. 1986. Introduction: The Problem of the Problematica. In A. Hoffman and M. H. Nitecki, eds., *Problematic Fossil Taxa*, pp. 3–11. Oxford: Oxford University Press.

Bengtson, S. and S. Conway Morris. 1984. A comparative study of Lower Cambrian *Halkieria* and Middle Cambrian *Wiwaxia*. *Lethaia* 17:307–329.

Bengtson, S., S. C. Matthews, and V. V. Missarzhevsky. 1986. The Cambrian netlike fossil *Microdictyon*. In A. Hoffman and M. H. Nitecki, eds., *Problematic Fossil Taxa*, pp. 97–115. Oxford: Oxford University Press.

Bengtson, S. and V. V. Missarzhevsky. 1981. Coeloscleritophora—a major group of enigmatic Cambrian metazoans. *United States Geological Survey Open-File Report* 81-743:19–21.

Berkner, L. V. and L. C. Marshall. 1965. History of major atmospheric components. *Proceedings of the National Academy of Sciences (USA)* 53:1215–1225.

Berner, R. A. and G. P. Landis. 1987. Chemical analysis of gaseous bubble inclusions in amber: The composition of ancient air? *American Journal of Science* 287:757–762.

Berner, R. A. and G. P. Landis. 1988. Gas bubbles in fossil amber as possible indicators of the major gas composition of ancient air. *Science* 239:1406–1409.

Bjorlykke, K. 1982. Correlation of late Precambrian and early Paleozoic sequences by eustatic sea-level changes and the selection of the Precambrian-Cambrian boundary. *Precambrian Research* 17:99–104.

Bloeser, B. 1985. *Melanocyrillium*, a new genus of structurally complex late Precambrian microfossils from the Kwagant Formation (Chuar Group), Grand Canyon, Ariz., *Journal of Paleontology* 59:741–765.

Bokuniewicz, H. J. and P. B. Gordon. 1980. Sediment transport and deposition in Long Island Sound. *Advances in Geophysics* 22:69–106.

Bond, G. C., N. Christie-Blick, M. A. Kominz and W. J. Devlin. 1985. An early Cambrian rift to post-rift transition in the Cordillera of western North America. *Nature* 315:742–745.

Brand, S., C. Sagan and L. Margulis. 1977. Controversy is rife on Mars. In S. Brand, ed., *Space Colonies*, pp. 120–127. San Francisco: Waller Press.

Brasier, M. D. 1979. The Cambrian radiation event. In M. R. House, ed., *The Origin of Major Invertebrate Groups*, pp. 103–159. London: Academic Press.

Brasier, M. D. 1986. Why do lower plants and animals biomineralize? *Paleobiology* 12:241–250.

Briand, F. and J. E. Cohen. 1987. Environmental correlates of food chain length. *Science* 238:956–960.

Briggs, D. E. G., and D. Collins. 1988. A Middle Cambrian chelicerate from Mount Stephen, British Columbia. *Palaeontology* 31:779–798.

Brøgger, W. C. 1886. Om alderen af *Olenellus* zonen i Nordiamerika. *Geoloiska Foereningen i Stockholm, Forhandlinger* 8:182–213.

Calvin, W. H. 1986. *The River That Flows Uphill.* San Francisco: Sierra Club Books.

Chen Jun-yuan and Qi Dun-lu. 1981. Upper Cambrian cephalopods from western Zhejiang. *Geological Society of America Special Paper* 187:137–141.

Chlupáč, I. and V. Havlíček. 1965. *Kodymirus* n.g., a new aglaspid merostome of the Cambrian of Bohemia. *Sborník Geologických Věd, řada P.* 6:7–27.

Cisne, J. L. 1975. Anatomy of *Triartrus* and the relationships of the Trilobita. *Fossils and Strata* 4:45–63.

Clarkson, E. K. N. 1979. *Invertebrate Paleontology and Evolution.* London: George Allen and Unwin.

Cloud, P. 1968. Pre-metazoan evolution and the origins of the Metazoa. In E. T. Drake, ed., *Evolution and Environment*, pp. 1–72. New Haven: Yale University Press.

Cloud, P. 1973. Pseudofossils: A plea for caution. *Geology* 1:123–127.

Cloud, P. 1976. Beginnings of biospheric evolution and their biogeochemical consequences. *Paleobiology* 2:351–387.

Cloud, P. 1983. The Biosphere. *Scientific American* 249:176–187.

Cloud, P. 1986. Reflections on the beginning of metazoan evolution. *Precambrian Research* 31:405–408.

Cloud, P. and M. F. Glaessner. 1982. The Ediacaran Period and System: Metazoa inherit the earth. *Science* 217:783–792.

Cloud, P. E. and C. A. Nelson. 1966. Phanerozoic-Cryptozoic and related transitions: New evidence. *Science* 154:766–770.

Cloud, P., S. M. Awramik, K. Morrison and D. G. Hadley. 1979. Earliest Phanerozoic or latest Proterozoic fossils from the Arabian Shield. *Precambrian Research* 10:73–93.

REFERENCES

Collins, D. 1985. A new Burgess shale type fauna in the Middle Cambrian Stephen Formation on Mt. Stephen, British Columbia. *Geological Society of America Abstracts with Program* 17:550.

Conway Morris, S. 1977. Fossil priapulid worms. *Special Papers in Palaeontology* 20:1–159.

Conway Morris, S. 1985. The Middle Cambrian metazoan *Wiwaxia corrugata* (Matthew) from the Burgess Shale and *Ogygopsis* Shale, British Columbia, Canada. *Philosophical Transactions of the Royal Society of London* B307:507–582.

Conway Morris, S. 1987a. The search for the Precambrian-Cambrian boundary. *American Scientist* 75:157–167.

Conway Morris, S. 1987b. Precambrian ancestors. *American Scientist* 75:570.

Conway Morris, S. 1988. Radiometric dating of the Precambrian-Cambrian boundary in the Avalon Zone. *New York State Museum Bulletin* 463:53–58.

Conway Morris, S. 1989. South-eastern Newfoundland and adjacent areas (Avalon Zone). In J. W. Cowie and M. D. Brasier, eds., *The Precambrian-Cambrian Boundary*, pp. 7–39. Oxford: Clarendon Press.

Conway Morris, S. and R. J. F. Jenkins. 1985. Healed injuries in Early Cambrian trilobites from South Australia. *Alcheringa* 9:167–178.

Cook, P. J. and M. W. McElhinny. 1979. A reevaluation of the spatial and temporal distribution of sedimentary phosphate deposits in the light of plate tectonics. *Economic Geology* 74:315–330.

Cook, P. J. and J. H. Shergold. 1984. Phosphorus, phosphorites, and skeletal evolution at the Precambrian-Cambrian boundary. *Nature* 308:231–236.

Cowen R. 1976. *History of Life*. New York: McGraw-Hill.

Cowen, R. 1986. The role of algal symbiosis in reefs through time. *North American Paleontological Convention, Abstracts with Program* 4:A10.

Crimes, T. P. 1987. Trace fossils and the Precambrian-Cambrian boundary. *Geological Magazine* 124:97–119.

Crimes, T. P. and J. D. Crossley. 1980. Inter-turbidite bottom current orientation from trace fossils with an example from the Silurian Flysch of Wales. *Journal of Sedimentary Petrology* 50:821–830.

Dalrymple, R. W., G. M. Narbonne and L. Smith. 1985. Eolian action and the distribution of Cambrian shales in North America. *Geology* 13:607–610.

Darwin, C. 1859 (6th edition, 1872). *On the Origin of Species by Means of Natural Selection*. London: John Murray.

Dawson, J. W. Möbius on Eozoön Canadense. *American Journal of Science* 16:196–197.

Debrenne, F. 1964. Archaeocyatha: Contribution à l'étude des faunes Cambriennes du Maroc, de Sardaigne et de France. *Serv. Mines Carte Géol. Maroc, Nôtes et Mémoires*, 1 and 2:1–265.

Debrenne, F. and P. D. Kruse. 1986. Shackleton limestone archaeocyaths. *Alcheringa* 10:235–278.

Delaca, T. E., D. M. Karl, and J. H. Lipps. 1981. Direct use of dissolved organic carbon by agglutinated benthic foraminifera. *Nature* 289:287–289.

Derstler, K. 1982. Helicoplacoids reinterpreted as triradiate edrioasteroids. *Geological Society of America Abstracts with Program* 14:159.

Dewey, J. F. and K. C. A. Burke. 1973. Plume-generated triple junctions: Key indicators in applying plate tectonics to old rocks. *Journal of Geology* 86:406–433.

Donovan, S. K. 1987. The fit of the continents in the late Precambrian. *Nature* 327:130–141.

Droser, M. L. and D. J. Bottjer. 1988. Trends in depth and extent of bioturbation in Cambrian carbonate marine environments, western United States. *Geology* 16:233–236.

Durham, J. W. 1971. The fossil record and the origin of the deuterostomata. *Proceedings of the North American Paleontological Convention* 1969H:1104–1132.

Durham, J. W. and K. E. Caster. 1963. Helicoplacoidea: a new class of echinoderms. *Science* 140:820–822.

Emmons, E. 1847. *Agriculture of New-York: Comprising an Account of the Classification, Composition and Distribution of the Soils and Rocks, and the Natural Waters of the Different Geological Formations; Together with a Condensed View of the Climate and the Agricultural Productions of the State.* Volume 1. Albany: C. Van Benthuysen.

Emmons, E. 1856. *American Geology, Containing a Statement of the Principles of the Science with Full Illustrations of the Characteristic American Fossils. With an Atlas and a Geological Map of the United States.* Volume 1. Albany: Sprague.

Erwin, D. H., J. W. Valentine, and J. J. Sepkoski, Jr. 1987. A comparative study of diversification events: The early Paleozoic versus the Mesozoic. *Evolution* 41:1177–1186.

Evans, J. W. 1910. The sudden appearance of the Cambrian fauna. *Eleventh International Geological Congress, Stockholm, 1910, Compte Rendu* 1:543–546.

Evitt, W. R. 1986. *Sporopollenin Dinoflagellate Cysts: Their Morphology and Interpretation.* Dallas, Texas: American Association of Stratgraphic Palynologists Foundation.

Fedonkin, M. A. 1978. Drevneishie iskopaemye sledy i puti evoluutsii povedeniya gryntoedov (Ancient trace fossils and the behavioral evolution of mud-eaters) *Paleontologicheskii Zhurnal* 2:106–111.

Fedonkin, M. A. 1981. *Belomorskaya biota venda: Dokembriiskaya besskeletnaya fauna severa Russkoi platformy* (White Sea biota of the Vendian: Precambrian nonskeletal fauna of the northern Russian Platform). Trudy Akademiya Nauk SSSR 342:1–100.

Fedonkin, M. A. 1987. *Besskeletnaya fauna venda i ee mesto v evolyutsii metazoa* (The nonskeletal fauna of the Vendian and its place in the evolution of metazoa). Akademiya Nauk SSSR, Trudy Paleontologiches-kogo Instituta, Tom 226. Moscow: Izdatel'stvo "Nauka".

Felbreck, H. 1981. Chemautotrophic potential of the hydrothermal vent tube worm, *Riftia pachyptila* Jones (Vestimentifera). *Science* 213:336–338.

Fischer, A. L. 1965. Fossils, early life, and atmospheric history. *Proceedings of the National Academy of Sciences (USA)* 53:1205–1213.

Fischer, A. L. 1984. Biological innovations and the sedimentary record. In H. D. Holland and A. F. Trendall, eds., *Patterns of Change in Earth Evolution*, p. 145–157. Berlin: Springer-Verlag.

Gehling, J. G. 1986. Algal binding of siliciclastic sediments: A mechanism in the preservation of Ediacaran fossils. *12th International Sedimentological Congress, Abstracts* 12:117.

Gehling, J. G. 1987. Earliest known echinoderm—a new Ediacaran fossil from the Pound Subgroup of South Australia. *Alcheringa* 11:337–345.

Gehling, J. G. 1988. A cnidarian of actinian-grade from the Ediacaran Pound Subgroup, South Australia. *Alcheringa* 12:299–314.

Germs, G. J. B. 1972. New shelly fossils from the Nama Group, South West Africa. *American Journal of Science* 272:752–761.

Germs, G. J. B., A. H. Knoll, and G. Vidal. 1986. Latest Proterozoic microfossils from the Nama Group, Namibia (South West Africa). *Precambrian Research* 32:45–62.

Geyer, G. 1988. Complementary notes on the Precambrian-Cambrian transition in Moroco. *New York State Museum Bulletin* 463:12–13.

Gibson, G. G., S. A. Teeter, and M. A. Fedonkin. 1984. Ediacaran fossils from the Carolina slate belt, Stanly County, North Carolina. *Geology* 12:387–390.

Gillett, S. L. 1985. The Cambrian explosion. *Amazing Science Fiction* 59(1):66–78.

Glaessner, M. F. 1959. Precambrian Coelenterata from Australia, Africa, and England. *Nature* 183:1472–1473.

Glaessner, M. F. 1969. Trace fossils from the Precambrian and basal Cambrian. *Lethaia* 2:369–393.

Glaessner, M. F. 1976. Early Phanerozoic annelid worms and their geological and biological significance. *Journal of the Geological Society of London* 132:259–275.

Glaessner, M. F. 1979. Precambrian. In R. A. Robinson and C. Teichert, eds., *Treatise on Invertebrate Paleontology*, Part A, pp. 79–118. Boulder, Colo. and Lawrence, Kansas: Geological Society of America and the University of Kansas.

Glaessner, M. F. 1984. *The Dawn of Animal Life: A Biohistorical Study.* Cambridge: Cambridge University Press.

Gleick, J. 1987. *Chaos: Making a New Science.* New York: Viking Penguin.

Gordon, P. 1987. Precambrian-Cambrian Boundary. *Origins Research* 10:7.

REFERENCES

Gould, S. J. 1980. The Panda's Thumb. New York: Norton.

Gould, S. J. 1984. The Ediacaran experiment. *Natural History* 93:14–23.

Gray, J. 1988. Evolution of the freshwater ecosystem: the fossil record. *Palaeography, Palaeoclimatology, Palaeoecology* 62:1–214.

Gürich, G. 1930. Die bislang ältesten Spuren von Organismen in Südafrika. *International Geological Congress, Comptes Rendus* 15:670–680.

Gürich, G. 1933. Die Kuibis-Fossilien der Nama-Formation von Südwestafrika. *Paläeontologische Zeitschrift* 15:137–154.

Hall, J. 1883. Cryptozoön N. G., *Cryptozoön proliferium* n. sp. *New York State Museum Annual Report* 36:1.

Hallock, P. 1981. Algal symbiosis: A mathematical analysis. *Marine Biology* 62:249–255.

Hallock, P. 1985. Drowned reefs and carbonate platforms: Reef-community disruption by nutrients provides a clue to the paradox. *Geological Society of America Abstracts with Program* 17:602.

Halstead Tarlo, L. B. 1967. *Xenusion*—onychophoran or coelenterate? *The Mercian Geologist* 2:97–99.

Hambrey, M. J. 1983. Correlation of Late Proterozic tillites in the North Atlantic region and Europe. *Geological Magazine* 120:309–320.

Häntzschel, W. 1975. Trace fossils and problematica. In R. A. Robinson and C. Teichert, eds., *Treatise on Invertebrate Paleontology*, part W., supplement 1. Boulder, Colo. and Lawrence, Kansas: Geological Society of America and the University of Kansas.

Harland, W. B. 1983. The Proterozoic glacial record. *Geological Society of America Memoir* 161:270–288.

Harrington, H. J. et al. 1959. Arthropoda. In R. C. Moore, ed., *Treatise on Invertebrate Paleontology*, part O, volume 1. New York: Geological Society of America and University of Kansas.

Hill, D. 1972. Archaeocyatha. In R. A. Robinson and C. Teichert, eds., *Treatise on Invertebrate Paleontology*, part E, volume 1. Boulder, Colo. and Lawrence, Kansas: Geological Society of America and the University of Kansas.

Hofmann, H. J. 1982. J. W. Dawson and 19th Century Precambrian paleontology. *Third North American Paleontological Convention, Proceedings* 1:243–242.

Hofmann, H. J. 1985. Precambrian carbonaceous megafossils. In D. F. Toomey and M. H. Nitecki, eds., *Paleoalgology*, pp. 20–33. Berlin: Springer-Verlag.

Hofmann, H. J. and M. P. Cecile. 1981. Occurrence of *Oldhamia* and other trace fossils in Lower Cambrian (?) argillites, Niddery Lake map area, Selwyn Mountains, Yukon Territory. *Geological Survey of Canada Paper* 81-1A:281–290.

Holser, W. T. 1977. Catastrophic chemical events in the history of the ocean. (*Nature* 267:403–408.)

REFERENCES

Hopfenberg, H. B. et al. 1988. Is the air in amber ancient? *Science* 241:717–724.

Howell, B. F. et al. 1944. Correlation of the Cambrian Formations of North America. *Geological Society of America Bulletin* 55:993–1003.

Hutchison, G. E. 1961. The biologist poses some problems. *American Association for the Advancement of Science Publications* 67:85–94.

Irving, E., R. F. Emslie, and H. Ueno. 1974. Upper Proterozoic poles from Laurentia and the history of the Grenville structural province. *Journal of Geophysical Research* 75:5491–5502.

Jaeger, H. and A. Martinsson. 1966. Remarks on the problematic fossil *Xenusion*. *Geologiska Foreningens i Stockholm Forhandlingar* 88:435–452.

Jenkins, R. J. F. 1984. Interpreting the oldest fossil cnidarians. *Palaeontographica Americana* 54:95–104.

Jenkins, R. J. F. 1985. The enigmatic Ediacaran (late Precambrian) genus *Rangea* and related forms. *Paleobiology* 11:336–355.

Jenkins, R. J. F. 1986. Are enigmatic markings in Adelaidean of Flinders Ranges fossil ice tracks? *Nature* 323:472.

Jenkins, R. J. F. 1988. Functional and ecological aspects of Ediacaran assemblages. In B. C. Klein-Helmuth and D. Savold, compilers, *American Association for the Advancement of Science Publication* 87–30, *Abstracts of Papers for the 1988 Annual Meeting, Boston*, p. 14. Washington, D.C.: American Association for the Advancement of Science.

Jenkins, R. J. F., C. H. Ford, and J. G. Gehling. 1983. The Ediacara Member of the Rawnsley Quartzite: the context of the Ediacara assemblage (Late Precambrian, Flinders Ranges). *Journal of the Geological Society of Australia* 105:101–119.

Jenkins, R. J. F. and J. G. Gehling. 1978. A review of the frond-like fossils of the Ediacara assemblage. *Records of the South Australian Museum* 17:347–359.

Johnson, M. E. 1982. The second geological career of Ebenezer Emmons: Success and Failure in the southern states, 1851–1860. In J.X. Corgan, ed., *The Geological Sciences in the Antebellum South*, pp. 142–171. University: University of Alabama Press.

Kaufmann, E. G. and F. Fursich. 1983. *Brooksella canyonensis:* a billion year old complex metazoan trace fossil from the Grand Canyon. *Geological Society of America Abstracts with Program* 15:608.

Kennard, J. M. and N. P. James. 1986. Thrombolites and stromatolites: Two distinct types of microbial structures. *Palaios* 1:492–503.

Kerr, R. A. 1987. Ancient air analyzed in dinosaur-age amber. *Science* 238:890.

Keto, L. S. and S. B. Jacobsen. 1985. The causes of 87Sr/86Sr variations in seawater of the past 750 million years. *Geological Society of America Abstracts with Program* 17:628.

Kidder, D. L., and K. Swett. 1989. Basal Cambrian reworked phosphates from Spitsbergen (Norway) and their implications. *Geological Magazine* 126:79–88.

Kirschvink, J. L., R. Kirk, and J. J. Sepkoski, Jr. 1982. Digital image enhancement of Ediacaran fossils: A first try. *Geological Society of America Abstracts with Program* 15:530.

Knoll, A. H. 1982. Microfossils frim the Late Precambrian Draken Conglomerate, Ny Friesland, Svalbard. *Journal of Paleontology* 56:755–790.

Knoll, A. H. 1985. A paleobiological perspective on sabkhas. In G. M. Friedman and W. E. Krumbein, eds., *Ecological Studies*, Vol. 53: Hypersaline Ecosystems, pp. 497–425. Berlin: Springer-Verlag.

Knoll, A. H. 1986. Algal fossils. *Science* 231:415.

Knoll, A. H., J. M. Hayes, A. J. Kaufman, K. Swett, and I. B. Lambert. 1986. Secular variation in carbon isotope ratios from Upper Proterozoic successions of Svalbard and East Greenland. *Nature* 321:832–838.

Knoll, A. H. and K. Swett. 1987. Micropaleontology across the Precambrian-Cambrian boundary in Spitsbergen. *Journal of Paleontology* 61:898–926.

Kohn, A. J. 1987. Progressive adaptation. *Science* 237:1235–1236.

Koneva, S. P. 1978. The first-record of ostracodes in the Lower Cambrian of Kazakhstan. *Paleontological Journal* 12:137–138.

Krumbiegel, G., H. Deichfuss, and H. Deichfuss. 1980. Ein neuer Fund von *Xenusion. Hallesches Jahrbuch fur Geowissenschaften* 5:97–99.

Kvenvolden, K. A. 1987. Methane hydrate: A major resevoir of carbon in the shallow geosphere. *Geological Society of America Abstracts with Program* 19:736.

Kvenvolden, K. A. and M. A. McMenamin. 1980. Hydrates of natural gas: A review of their geologic occurrence. *United States Geological Survey Circular* 825:1–11.

LaBarbera, M. 1981. Water flow patterns in and around three species of articulate brachiopods. *Journal of Experimental Marine Biology and Ecology* 55:185–206.

Lambert, I. B., M. R. Walter, Zang Wentong, Lu Songnian, and Ma Gougan. 1987. Palaeoenvironmental and carbon isotope stratigraphy of Upper Proterozoic carbonates of the Yangtze Platform. *Nature* 325:140–142.

Lapworth, C. 1879. On the tripartite classification of the Lower Palaeozoic rocks. *Geological Magazine* 6:1–15.

Lapworth, C. 1888. On the discovery of the *Olenellus* fauna in the Lower Cambrian rocks of Britian. *Nature* 39:212–213.

Landing, E. 1984. Skeleton of lapworthellids and the suprageneric classification of tommotiids (Early and Middle Cambrian phosphatic problematica). *Journal of Paleontology* 58:1380–1398.

Lewin, R. 1983. Alien beings here on earth? *Science* 223:39.

Lindsay, J. R., R. J. Korsch, and J. R. Wilford. 1987. Timing and breakup of a Proterozoic supercontinent: Evidence from Australian intacratonic basins. *Geology* 15:1061–1064.

Lochman-Balk, C. 1964. Paleo-ecologic studies of the Deadwood Formation (Cambro-Ordovician). *Proceedings of the 22nd International Geological Congress* 8:4–38.

REFERENCES

Lowenstam, H. A. 1980. What, if anything, happened at the transition from the Precambrian to the Phanerozoic? *Precambrian Research* 11:89–91.

Lyell, C. 1830. *Principles of Geology.* London: Murray.

Lyell, C. 1845. *Travels in North America.* London: Murray.

Lyell, C. 1849. *A Second Visit to North America.* London: Murray.

MacCracken, M. C. 1987. The chlorofluorocarbon dilemma. *Science* 238:598.

McElhinny, M. W. 1979. *Paleomagnetism and Plate Tectonics.* Cambridge: Cambridge University Press.

McMenamin, D. S., S. Kumar, and S. M. Awramik. 1983. Microbial fossils from the Kheinjua Formation, Middle Proterozoic Semri Group (Lower Vindhyan), Son Valley area, central India. *Precambrian Research* 21:247–271.

McMenamin, M. A. S. 1982. A case for two late Proterozoic-earliest Cambrian faunal province loci. *Geology* 10:290–292.

McMenamin, M. A. S. 1984. Paleontology and stratigraphy of Lower Cambrian and Upper Proterozoic sediments, Caborca Region, Northwestern Sonora, Mexico. Ph.D. diss., University of California at Santa Barbara.

McMenamin, M. A. S. 1985. Basal Cambrian small shelly fossils from the La Ciénega Formation, northwestern Sonora, Mexico. *Journal of Paleontology* 59:1414–1425.

McMenamin, M. A. S. 1986. The Garden of Ediacara. *Palaios* 1:178–182.

McMenamin, M. A. S. 1987a. The emergence of animals. *Scientific American* 256:94–102.

McMenamin, M. A. S. 1987b. Lower Cambrian trilobites, zonation, and correlation of the Puerto Blanco Formation, Sonora, Mexico. *Journal of Paleontology* 61:738–749.

McMenamin, M. A. S. 1988, Paleoecological feedback and the Vendian-Cambrian transition. *Trends in Ecology and Evolution* 3:205–208.

McMenamin, M. A. S. 1989. The origins and radiation of the early metazoa. In K. C. Allen and D. E. G. Briggs, eds., *Evolution of the Fossil Record,* pp. 73–98. London: Belhaven Press.

McMenamin, M. A. S., S. M. Awramik, and J. H. Stewart. 1983. Precambrian-Cambrian transition problem in western North America: part 2. Early Cambrian skeletonized fauna and associated fossils from Sonora, Mexico. *Geology* 11:227–230.

McMenamin, M. A. S. and D. S. McMenamin. 1987. Late Cretaceous atmospheric oxygen. *Science* 235:1561–1562.

McNamara, K. J. and D. M. Rudkin. 1984. Techniques of trilobite exuviation. *Lethaia* 17:153–173.

Mankiewicz, C. 1988. *Obruchevella* in the Middle Cambrian Burgess Shale: preservation and taxonomic affinity. *Geological Society of America Abstracts with Programs* 20:A226.

Margulis, L. 1981. *Symbiosis in Cell Evolution.* San Francisco: Freeman.

Margulis, L. and J. E. Lovelock. 1986. The atmosphere as circulatory system of the biosphere—the Gaia Hypothesis. In A. Kleiner and S. Brand, eds., *News That Stayed News,* pp. 15–25. Berkeley, Calif.: North Point Press.

Margulis, L. and K. V. Schwartz. 1982. *Five Kingdoms.* San Francisco: Freeman.

Matthew, G. F. 1900. Mr. Walcott's view of the Etcheminian. *The American Geologist* 25:255–258.

Matthews, S. C. and V. V. Missarzhevskii. 1975. Small shelly fossils of Late Precambrian and Early Cambrian age: a review of recent work. *Journal of the Geological Society of London* 131:289–304.

Mel'nikova, L. M. 1987. Bradoriidy iz tiskreskoi svity (nishnii kembrii) Estonii (Bradoriids from the Tiskres Suite of Estonia.) *Paleontologicheskii Zhurnal* 1:128–131.

Möbius, K. A. 1879. Professor Moebius on the Eozoön question. *Nature* 20:272.

Monastersky, R. 1987. Set adrift by wandering hotspots. *Science News* 132:250–252.

Morris, N. J. 1979. On the origin of Bivalvia. In M. R. House, ed., *The Origin of Major Invertebrate Groups,* pp. 381–413. London: Academic Press.

Mount, J. F. and S. M. Rowland. 1981. Grand Cycle A (Lower Cambrian) of the southern Great Basin: A product of differential rates of sea-level rise. In M. E. Taylor, ed., *Short Papers for the Second International Symposium on the Cambrian System,* pp. 143–146. Golden, Colo.: U.S. Geological Survey.

Müller, K. J. 1981. Arthropods with phosphatized soft parts from the Upper Cambrian "Orsten" of Sweden. In M. E. Taylor, ed., *Short Papers for the Second International Symposium on the Cambrian System,* pp. 147–157. Golden, Colo.: U.S. Geological Survey.

Narbonne, G. M. 1987. Trace fossils, small shelly fossils, and the Precambrian-Cambrian boundary. *Episodes* 10:339–340.

Narbonne, G. M. and H. J. Hofmann. 1987. Ediacaran biota of the Wernecke Mountains, Yukon, Canada. *Palaeontology* 30:647–676.

Narbonne, G. M., P. M. Myrow, E. Landing, and M. M. Anderson. 1987. A candidate stratotype for the Precambrian-Cambrian boundary, Fortune Head, Burin Peninsula, southeastern Newfoundland. *Canadian Journal of Earth Science* 24:1279–1293.

Nitecki, M. H. and F. Debrenne. 1979. The nature of radiocyathids and their relationship to receptaculitids and archaeocyathids. *Geobios* 12:5–27.

O'Brien, C. F. 1970. *Eozoön Canadense,* "The Dawn Animal of Canada." *Isis* 61:206–223.

O'Brien, C. F. 1971. On *Eozoön Canadense. Isis* 62:381–383.

Ogurtsova, R. N. and V. N. Sergeyev. 1987. Mikrobiota chichkanskoy svity verknego dokembriya Malogo Karatau (Yuzhnyy Kasakhstan) (The microbiota of the upper Precambrian Chichkanskaya Formation in the

REFERENCES

Lesser Karatau Region: Southern Kazakhstan). *Paleontologicheskii Zhurnal* 2:107–116.

Owen-Smith, N. 1987. Pleistocene extinctions: the pivotal role of megaherbivores. *Paleobiology* 13:351–362.

Paul, C. R. C. 1979. Early echinoderm radiation. In M. R. House, ed., *The Origin of Major Invertebrate Groups*, pp. 415–434. London: Academic Press.

Pearse, V., J. Pearse, M. Buchsbaum and R. Buchsbaum. 1987. *Living Invertebrates*. Palo Alto, California: Blackwell.

Piper, J. D. A. 1987. *Palaeomagnetism and the Continental Crust.* New York: Wiley.

Pojeta, J., Jr. 1981. Paleontology of Cambrian Mollusks. In M. E. Taylor, ed., *Short Papers for the Second International Symposium on the Cambrian System*, pp. 163–166. Golden, Colo.: U.S. Geological Survey.

Pojeta, J., Jr. 1987. Phylum Hyolitha. In R. S. Boardman, A. H. Cheetham, and A. J. Rowell, eds., *Fossil Invertebrates*, pp. 436–444. Palo Alto, Calif.: Blackwell Scientific Publications.

Pojeta, J., Jr., B. Runnegar, J. S. Peel, and M. Gordon, Jr. 1987. Phylum Mollusca. In R. S. Boardman, A. H. Cheetham, and A. J. Rowell, eds., *Fossil Invertebrates*, pp. 270–435. Palo Alto, Calif.: Blackwell Scientific Publications.

Pompeckj, J. F. 1927. Ein neues Zeugnis uralten Lebens. *Palaeontologische Zeitschrift* 9:287–313.

Poulsen, C. 1960. Notes on some Lower Cambrian fossils from French West Africa. *Matematisk-Fysiske Meddelelser udgivet af Det Kongelige Danske Videnskavernes Selskab* 32:1–20.

Resser, C. E. and B. F. Howell. 1938. Lower Cambrian *Olenellus* Zone of the Appalachians. *Geological Society of America Bulletin* 49:195–248.

Retallack, G. J. and C. R. Feakes. 1987. Trace fossil evidence for Late Ordovician animals on land. *Science* 235:61–63.

Rezack, R. 1957. *Girvanella* not a guide to the Cambrian. *Geological Society of America Bulletin* 68:1411–1412.

Rhoads, D. C. 1970. Mass properties, stability, and ecology of marine muds related to burrowing activity. In T. P. Crimes and J. C. Harper, eds., *Trace Fossils*, p. 391–406. Liverpool: Seel House Press.

Ricklefs, R. E. 1976. *The Economy of Nature.* Portland, Ore.: Chiron Press.

Riding, R. and L. Voronova. 1982. Calcified cyanophytes and the Precambrian-Cambrian transition. *Naturwissenschaften* 69:498–499.

Rivkin, R. B., I. Bosch, J. S. Pearse, and E. J. Lessard. 1986. Bacterivory: A novel feeding mode for asteroid larvae. *Science* 233:1311–1314.

Robison, R. A. 1965. Middle Cambrian eocrinoids from western North America. *Journal of Paleontology* 39:355–364.

Robison, R. A. 1987. Superclass Trilobitomorpha. In R. S. Boardman, A. H. Cheetham, and A. J. Rowell, eds., *Fossil Invertebrates*, pp. 221–241. Palo Alto, Calif.: Blackwell Scientific Publications.

Robison, R. A. and R. L. Kaesler. 1987. Phylum Arthropoda. In R. S. Boardman, A. H. Cheetham, and A. J. Rowell, eds., *Fossil Invertebrates*, pp. 205–221. Palo Alto, Calif.: Blackwell Scientific Publications.

Rowell, A. J. 1977. Early Cambrian brachiopods from the southwestern Great Basin of California and Nevada. *Journal of Paleontology* 51:68–85.

Rowell, A. J. and N. E. Caruso. 1985. The evolutionary significance of *Nisusia sulcata*, an early articulate brachiopod. *Journal of Paleontology* 59:1227–1242.

Rowland, S. 1983. A new shirt for Carl. *Science 83* 4(4):80–82.

Rowland, S. M. 1984. Were there framework reefs in the Cambrian? *Geology* 12:181–183.

Rowland, S. M. 1988. Archaeocyatha: Cambrian relicts of the Ediacaran? In B. C. Klein-Helmuth and D. Savold, compilers, *American Association for the Advancement of Science Publication 87–30, Abstracts of Papers for the 1988 Annual Meeting, Boston, Program Changes*, p. 6. Washington, D.C.: American Association for the Advancement of Science.

Rozanov, A. Yu., V. V. Missarzhevskij, N. A. Volkova, L. G. Voronova, I. N. Krylov, B. M. Keller, I. K. Korolyuk, K. Lendzion, R. Michniak, N. G. Pychova, and A. D. Sidorov. 1969. *Tommotskii Yarus i Problema Nishnei Granitsy Kembriya* (Tommotian Stage and the Cambrian Lower boundary problem). Akademiya Nauk SSSR, Trudy Geologicheskii Institut 206:1–380. (English edition, 1981, New Delhi: Amerind.)

Rudkin, D. M. 1979. Healed injuries in *Ogygopsis klotzi* (trilobite) from the Middle Cambrian of British Columbia. *Royal Ontario Museum of Life Sciences Occasional Paper* 32:1–8.

Rudman, W. B. 1987. Solar-powered animals. *Natural History* 96:50–53.

Runnegar, B. 1982. Oxygen requirements, biology, and phylogenetic significance of the Late Precambrian worm *Dickinsonia*, and the evolution of the burrowing habit. *Alcheringa* 6:223–239.

Runnegar, B. and C. Bentley. 1983. Anatomy, ecology, and affinities of the Australian Early Cambrian bivalve *Pojetaia runnegari* Jell. *Journal of Paleontology* 57:73–92.

Runnegar, B., J. Pojeta, Jr., M. E. Taylor, and D. Collins. 1979. New species of the Cambrian and Ordovician chitons *Matthevia* and *Chelodes* from Wisconsin and Queensland: Evidence for the early history of the polyplacophoran mollusks. *Journal of Paleontology* 53:1374–1394.

Russell, D. A. 1981. Speculations on the evolution of intelligence in multicellular organisms. In J. Billingham, ed., *Life in the Universe*, pp. 259–275. Cambridge, Mass.: MIT Press.

Sabrodin, W. 1972. Leben im Prakambrium. *Bild der Wissenschaft* 1972:586–591.

Salvini-Plawen, L. V. and E. Mayr. 1977. On the evolution of photoreceptors and eyes. *Evolutionary Biology* 10:207–263.

St. Jean, J. 1972. A new Cambrian trilobite from the Piedmont of North Carolina. *American Journal of Science* 273-A:196–216.

REFERENCES

Sawkins, F. J. 1976. Widespread continental rifting: Some considerations of timing and mechanism. *Geology* 4:427–430.

Schallreuter, R. 1985. Das zweite *Xenusion*. *Geschiebekunde Aktuell* 1(2):19–23.

Schneer, C. J. 1969. Ebenezer Emmons and the foundations of American geology. *Isis* 60:439–450.

Schneer, C. J. 1978. The great Taconic controversy. *Isis* 69:173–191.

Schopf, J. W. 1981. The Precambrian development of an oxygen atmosphere. *United States Geological Survey Professional Paper* 1161:B1–B11.

Schuchert, C. and C. O. Dunbar. 1933. *Historical Geology*. New York: Wiley.

Sears, J. W. and R. A. Price. 1978. The Siberian connection: A case for Precambrian separation of the North American and Siberian cratons. *Geology* 6:267–270.

Secord, J. A. 1986. *Controversy in Victorian Geology: The Cambrian-Silurian Dispute*. Princeton: Princeton University Press.

Sedgwick, A. and R. I. Murchison. 1835. On the *Silurian* and *Cambrian Systems*, exhibiting the order in which the older Sedimentary Strata succeed each other in England and Wales. *Philosophical Magazine* 3(7):483–485.

Seilacher, A. 1956. Der Beginn des Kambriums als biologische Wende. *Neues Jb. Geol. u. Paläontol., Abh.* 103(½):155–180.

Seilacher, A. 1957. An-aktualistiches Wattenmeer. *Paläeotologische Zeitschrift* 31:198–206.

Seilacher, A. 1972. Divaricate patterns in pelecypod shells. *Lethaia* 5:325–343.

Seilacher, A. 1977. Evolution of trace fossil communities. In A. Hallam, ed., *Patterns of Evolution as Illustrated by the Fossil Record*, pp. 357–376. Amsterdam: Elsevier.

Seilacher, A. 1983. Precambrian metazoan extinctions. *Geological Society of America Abstracts with Program* 15:683.

Seilacher, A. 1984. Late Precambrian and Early Cambrian Metazoa: preservational or real extinctions? In H. D. Holland and A. F. Trendall, eds., *Patterns of Change in Earth Evolution*, pp. 159–168. Berlin: Springer-Verlag.

Seilacher, A. 1985. Discussion of Precambrian metazoans. *Philosophical Transactions of the Royal Society of London* B311:47–48.

Seilacher, A. 1985. Trilobite palaeobiology and substrate relationships. *Transactions of the Royal Society of Edinburgh* 76:231–237.

Seilacher, A. 1986. Evolution of behavior as expressed in marine trace fossils. In M. H. Nitecki and J. Kitchell, eds., *Evolution of Behavior*, pp. 62–87. Oxford: Oxford University Press.

Seilacher, A., W.-E. Reif, and F. Westphal. 1985. Sedimentological, ecological, and temporal patterns of fossil Lagerstatten. *Philosophical Transactions of the Royal Society of London* B311:5–23.

Sepkoski, J. J., Jr. 1982. Flat-pebble conglomerates, storm deposits, and the Cambrian bottom fauna. In G. Einsele and A. Seilacher, eds., *Cyclic and Event Stratification*, pp. 371–385. Berlin: Springer-Verlag.

Shear, W. A., P. M. Bonamo, J. D. Grierson, W. D. Ian Rolfe, E. L. Smith, and R. A. Norton. 1984. Early Land Animals in North America: Evidence from Devonian Age Arthropods from Gilboa, New York. *Science* 224:492–494.

Signor, P. W. and M. A. S. McMenamin. 1988. The Early Cambrian worm tube *Onuphionella* from California and Nevada. *Journal of Paleontology* 62:233–240.

Signor, P. W., P. M. Sadler, and J. F. Mount. 1988. Stratigraphic completeness and the limits of biostratigraphic correlation in earliest Cambrian sediments. *New York State Museum Bulletin* 463:16–17.

Sokolov, B. S. 1952. O vozraste drevneishego osadochnogo nokrova Russkoi platformy (On the age of the oldest sedimentary cover of the Russian platform). *Izdatel'stvo Akademiya Nauka SSSR, Seriya Geol.* 5:21–31.

Sokolov, B. S. and M. A. Fedonkin. 1984. The Vendian as the terminal system of the Precambrian. *Episodes* 7:12–19.

Sokolov, B. S. and M. A. Fedonkin. 1986. Global biological events in the late Precambrian. In S. Bhattacharji, G. M. Friedman, H. J. Neugebauer, and A. Seilacher, eds., *Global Bio-Events*, pp. 105–107. Berlin: Springer-Verlag.

Sokolov, B. S. and A. B. Ivanovskii, eds. 1985. *Vendskaya Sistema. Istoriko-geologicheskoe i paleontologicheskoe obosnovanie. T. 1. Paleontologiya.* (The Vendian System. Historical-Geological and Paleontological Bases. Volume 1. Paleontology.) Moscow: Izdatel'stvo "Nauka".

Sprigg, R. C. 1947. Early Cambrian "Jellyfishes" of Ediacara, South Australia, and Mount John, Kimberly District, Western Australia. *Transactions of the Royal Society of South Australia* 71:212–224.

Sprinkle, J. and P. M. Kier. 1987. Phylum Echinodermata. In R. S. Boardman, A. H. Cheetham, and A. J. Rowell, eds., *Fossil Invertebrates*, pp. 550–611. Palo Alto, Calif.: Blackwell Scientific Publications.

Stanley, S. M. 1976. Fossil data and the Precambrian-Cambrian evolutionary transition. *American Journal of Science* 276:56–76.

Steele, J. H. 1825. A description of the Oolite Formation lately discovered in the county of Saratoga and in the state of New York. *American Journal of Science* 9:16–19.

Steneck, R. S. 1983. Escalating herbivory and resulting adaptive trends in calcareous algal crusts. *Paleobiology* 9:44–61.

Steneck, R. S. 1985. Adaptations of crustose coralline algae to herbivory: Patterns in space and time. In D. F. Toomey and M. H. Nitecki, eds., *Paleoalgology*, pp. 352–366. Berlin: Springer-Verlag.

Stewart, J. H. 1976. Late Precambrian evolution of North America: plate tectonic implication. *Geology* 4:11–15.

REFERENCES

Stewart, J. H. and C. A. Suczek. 1977. Cambrian and latest Precambrian paleogeography and tectonics in the western United States. *Society of Economic Paleontologists and Mineralogists, Pacific Coast Paleogeography Symposium* 1:1–17.

Stewart, J. H., M. A. S. McMenamin, and J. M. Morales. 1984. Upper Proterozoic and Cambrian Rocks in the Caborca Region, Sonora, Mexico— Physical Stratigraphy, Biostratigraphy, Paleocurrent Studies and Regional Relations. *United States Geological Survey Professional Paper* 1309.

Stubblefield, C. J. 1956. Cambrian Paleogeography in Britian. *XX Congreso Geologico Internacional, México, 1956. El sistema Cambrico su Paleogeografia y el problema de su base* 1:1–43.

Sun Weiguo. 1986. Precambrian medusoids: the *Cyclomedusa*-plexus and *Cyclomedusa*-like pseudofossils. *Precambrian Research* 31:325–360.

Sun Weiguo, Wang Guixiang, and Zhou Benhe. 1986. Macroscopic wormlike body fossils from the Upper Precambrian (900–700 Ma), Huainan District, Anhui, China, and their stratigraphic and evolutionary significance. *Precambrian Research* 31:377–403.

Szaniawski, H. 1982. Chaetognath grasping spines recognized among Cambrian protoconodonts. *Journal of Paleontology* 56:860–810.

Thayer, C. 1977. Biological bulldozers and the evolution of marine benthic communities. *Science* 203:458–461.

Towe, K. M. 1981. Biochemical keys to the emergence of complex life. In J. Billingham, ed., *Life in the Universe*, pp. 297–89. Cambridge, Mass.: MIT Press.

Tyler, S. A. and E. S. Barghoorn. 1954. Occurrence of structurally preserved plants in pre-Cambrian rocks of the Canadian Shield. *Science* 119:606–608.

Ushatinskaya, G. T. 1986. Nakhodka drevneyshevey zamkovoy brakhiopody (The discovery of the oldest articulate brachiopods). *Paleontologicheskii Zhurnal* 4:102–103.

Valentine, J. W. 1969. Patterns of taxonomic and ecologic structure of the shelf benthos during Phanerozoic time. *Palaeontology* 12:684–709.

Valentine, J. W. 1973. *Evolutionary Paleoecology of the Marine Biosphere.* Englewood Cliffs, N.J.: Prentice-Hall.

Valentine, J. W. 1986. Fossil record of the origin of bauplane and its implications. In D. M. Raup and D. Jablonski, eds., *Patterns and Processes in the History of Life*, pp. 209–222. Berlin: Springer-Verlag.

Valentine, J. W. and D. H. Erwin. 1987. Interpreting great developmental experiments: The fossil record. In R. A. Raff and E. C. Raff, eds., *Development as an Evolutionary Process*, pp. 71–107. New York: Alan R. Liss.

Valentine, J. W. and E. M. Moores. 1972. Global tectonics and the fossil record. *Journal of Geology* 80:167–184.

Vermeij, G. J. 1987. *Evolution and Escalation.* Princeton: Princeton University Press.

Vidal, G. 1976. Late Precambrian microfossils from the Visingsö Beds in southern Sweden. *Fossils and Strata* 9:1–157.

Vidal, G. 1984. The oldest eukaryotic cells. *Scientific American* 250:48–57.

Vidal, G. and A. H. Knoll. 1982. Radiations and extinctions of plankton in the late Proterozoic and early Cambrian. *Nature* 296:57–60.

Wainright, S. C. 1987. Stimulation of heterotrophic microplankton production by resuspended marine sediments. *Science* 238:1710–1712.

Walcott, C. D. 1887. Fauna of the "Upper Taconic" of Emmons, in Washington County, N. Y. *American Journal of Science* 34:187–199.

Walcott, C. D. 1889. Stratigraphic position of the *Olenellus* Fauna in North America and Europe. *American Journal of Science* 37:374–392.

Walcott, C. D. 1890. Stratigraphic position of the *Olenellus* Fauna in North America and Europe (continued). *American Journal of Science* 38:29–42.

Walcott, C. D. 1891. Correlation papers—Cambrian. *Bulletin of the U.S. Geological Survey* 81:1–447.

Walcott, C. D. 1899. Pre-Cambrian fossiliferous formations. *Bulletin of the Geological Society of America* 10:199–214.

Walcott, C. D. 1900. Lower Cambrian terrane in the Atlantic province. *Proceedings of the Washington Academy of Sciences* 1:301–339.

Walcott, C. D. 1910. Cambrian Geology and Paleontology 2. No. 1, Abrupt appearance of the Cambrian fauna on the North American continent. *Smithsonian Miscellaneous Collections* 57:1–16.

Walcott, C. D. 1912. Cambrian Geology and Paleontology 2. No. 9, New York Potsdam-Hoyt fauna. *Smithsonian Miscellaneous Collections* 57:251–305.

Walcott, C. D. 1914. Pre-Cambrian Algonkian algal flora. *Smithsonian Miscellaneous Collections* 64:1–75.

Walcott, C. D. 1920. Cambrian Geology and Paleontology IV. No. 6, Middle Cambrian Spongiae. *Smithsonian Miscellaneous Collections* 67:475–556.

Walker, J. C. G. 1987. Was the Archaean biosphere upside down? *Nature* 329:710–712.

Walter, M. R. 1972. Tectonically deformed sand volcanoes in a Precambrian Greywacke, Northern Territory of Australia. *Journal of the Geological Society of Australia* 18:395–399.

Walter, M. R. 1976. Hot spring sediments in Yellowstone National Park. In M. R. Walter, ed., *Stromatolites*, pp. 489–498. Amsterdam: Elsevier.

Watson, A. J., J. E. Lovelock, and L. Margulis. 1978. Methanogenesis, fires, and the regulation of atmospheric oxygen. *BioSystems* 10:293–298.

Wegener, A. 1967. *The Origin of Continents and Oceans.* English translation of 4th ed. [1929]. London: Methuen.

REFERENCES

Weisburd, S. 1986. Hydrothermal discoveries from the deep. *Science News* 130:389.

Whittington, H. B. 1985. *The Burgess Shale.* New Haven: Yale University Press.

Whittington, H. B. and D. E. Briggs. 1985. The largest Cambrian animal, *Anomalocaris,* Burgess Shale, British Columbia. *Philosophical Transactions of the Royal Society of London* B309:569–609.

Wilson, J. T. 1966. Did the Atlantic close and then reopen? *Nature* 211:676–679.

Wilson, M. E. 1931. Life in the pre-Cambrian of the Canadian Shield. *Transactions of the Royal Society of Canada* 25:119–126.

Yankauskas, T. V. 1978. Riphean plant microfossils from the Southern Urals. *Transactions of the USSR Academy of Sciences* 242:98–100.

Yochelson, E. L. 1979. Early radiation of Mollusca and Mollusc-like groups. In M. R. House, ed., *The Origin of Major Invertebrate Groups,* pp. 323–358.

Young, G. M. 1982. The late Proterozoic Tindir Group, east-central Alaska: evolution of a continental margin. *Geological Society of America Bulletin* 93:759–783.

Zaine, M. F. and T. R. Fairchild. 1987. Novas considerações sobre os fósseis da Formação Tamengo, Grupo Corumbá, S W do Brasil. *Congresso Brasileiro de Paleontologia, Anais* 10:797–807.

Zonenshain, L. P., M. I. Kuzmin, and M. V. Kononov. 1985. Absolute reconstructions of the Paleozoic oceans. *Earth and Planetary Science Letters* 74:103–116.

INDEX

INDEX